STRIKE
AT PILKINGTONS

*Tony Lane
and Kenneth Roberts*

COLLINS/FONTANA

First published in Fontana 1971
© Tony Lane and Kenneth Roberts, 1971

Printed in Great Britain
for the publishers William Collins Sons & Co Ltd,
14 St James's Place, London, S.W.1,
by Richard Clay (The Chaucer Press), Ltd,
Bungay, Suffolk

CONDITIONS OF SALE
This book is sold subject to the condition
that it shall not, by way of trade or otherwise,
be lent, re-sold, hired out or otherwise circulated
without the publisher's prior consent in any form of
binding or cover other than that in which it is
published and without a similar condition
including this condition being imposed
on the subsequent purchaser

CONTENTS

Introduction   11

PART I

1. *The Social, Economic, and Political Background*   25
   St Helens   25
   The glass empire of Pilkington Brothers   33
   Life and labour on the shop-floor   44
   The General and Municipal Workers Union: an empire of labour   49
   The trade unions and the power structure   56
2. *A Strike Diary, and the Role of the Press*   64
   Strike diary   64
   What some of the papers said   71
   The press and the 'public'   76
   The press and the 'generals'   80

PART II

3. *The Rank and File*   85
   The unknown but decisive factor   85
   The outbreak of the strike   86
   The union and the Rank and File Strike Committee   96
   The invisible majority   99
   The politics of the rank and file striker   102
   The experience of the rank and file striker   104
   The influence of the rank and file   107
4. *The Union*   109
   91 branch   110
   The Joint Industrial Council   114
   The calm before the storm   117
   An equivocal beginning   119
   Mass humiliation   124

CONTENTS

The £3 offer	127
The opposing forces	130
'Victory'	134
5. *Pilkingtons*	138
The Pilkington 'style'	139
Gunpowder + engineering = strike	143
Playing it 'cool'	147
The 'unofficial people'	153
'Good and reliable men'	154
Climax, anti-climax, climax	156
6. *The Rank and File Strike Committee (I)*	158
The emergence of the Rank and File Strike Committee	158
What sort of men?	163
Strike organisation	167
Picketing	170
'Blackings'	171
Mass meetings	173
Delegations	174
'Outside influences'	176
7. *The Rank and File Strike Committee (II)*	180
An element of surprise	184
A 'powder keg'?	185
The question of paternalism	187
Civil war in the GMWU	190
Pilkingtons, the police, and the press	195
The magic words – 'The Trades Union Congress'	198
'By God, this strike has been a bloody education!'	201

PART III

8. *Aftermath*	207
The Trades Union Congress – not so magical after all	207
Pilkingtons and the GMWU	209
The Court of Inquiry	212
The Glass and General Workers Union	215
The problems of the breakaway union	220
9. *Conclusions*	223
Strikes are 'normal'	223

CONTENTS

An industrial protest movement?	227
A strike-conscious culture	232
The escalation of conflict	235
The prospects of peace	240
Appendix	245
Selected documents	247

ACKNOWLEDGEMENTS

Thanks are due to many people, but they are not too numerous to mention. Strikes are not easy to study at the close range at which we attempted to work – as those who try to follow us will discover for themselves. While we undoubtedly had more in the way of luck than we could reasonably have expected, we also enjoyed the confidence, the hospitality, the co-operation, and the completely open friendliness of large numbers of people. To them our very sincere thanks and our hope that none of them will feel that we have betrayed their confidence and friendship.

Our thanks first to all those anonymous people in St Helens, the rank and file strikers and their families. Secondly, to all those who played prominent parts in the strike: David Basnett, George Bibby, Bill Bradburn, Jimmy Beech, Alan Brennan, Derek Bridge, Gerry Caughey, Mike Considine, Bill Cowley, Harry Erlam, Derek Greenough, Denis Hallissy, George Crooks, Jimmy Crosby, Jimmy Hevey, Terry Hodgson, Harold Hunt, Ron Jones, Don Kenny, Gordon Lawson, Joe Leonard, Bill Measures, Derek Pearce, Jim Peers, John Potter, John Riley, Joe Robinson, Mick Swain, Stan Swift, Tommy Wall, Martin Walsh, Johnny Wilson, Brian Willey, and F. L. Wells. Our thanks also to the board of directors and the public relations department of Pilkington Brothers Ltd.

To all our *friends* of the press, particularly David Wilson of *The Financial Times*, Graham Nown of *The Lancashire Evening Post & Chronicle*, and the library staff of *The Guardian* in Manchester.

To our colleagues in the Department of Social Science, University of Liverpool, Professor J. B. Mays, Stan Clarke, Patrick McNabb, Nick Kokosolakis, Garry Littlejohn, Gideon Ben-Tovim who helped us in many ways, some of which they will not have been aware. To John Harris of the Department of Economics who told us most of what there is to know about St Helens. To David Purdy of the Department

ACKNOWLEDGEMENTS

of Economics, Manchester University, who, at very short notice, produced an excellent analysis of the history and economic performance of Pilkingtons, our apologies for being able to use so little of what he provided.

To Diana Lane and Janette Harper who took on the tedious chore of transcribing and typing our tape-recorded interviews, and to Margaret D'Arcy, Hilary Farrell, Jennifer Geach, Barbara Kelsey, Irene Mercer, and Thea Wiltshire who typed our manuscript.

To Theo Nichols of the Department of Sociology at Bristol University who gave us good advice on the manuscript, and to Peter Shaw who checked some of it for intelligibility.

Finally the inevitable disclaimer – the responsibility for what lies in these pages is ours alone.

TONY LANE
KENNETH ROBERTS

LIST OF ABBREVIATIONS

RFSC	Rank and File Strike Committee
GMWU	General and Municipal Workers Union
T & GWU	Transport and General Workers Union
AEF	Amalgamated Engineers and Foundryworkers
TUC	Trades Union Congress
NEDC	National Economic Development Council
JIC	Joint Industrial Council
DEP	Department of Employment and Productivity
NEC	National Executive Committee of the GMWU
GGWU	Glass and General Workers Union

INTRODUCTION

For seven weeks in the Spring of 1970 the Pilkington glass-workers at St Helens, Lancashire, were out on unofficial strike, almost one hundred years to the month since the six-month strike in 1870 when the men returned on the employers' terms; a 15–23% reduction in pay. This time the men went back £3 a week to the good, although they had won that much after two weeks of the strike.

The strike started in the Flat Drawn department of the Sheet Works (one of the six separate Pilkington plants in St Helens) over an error in wage calculations, but within a matter of an hour or so it had been converted into a claim for 2/6d an hour on the basic rate of pay. By half past three on the afternoon of Friday, 3rd April, about 400 men had walked out. By half past three on Sunday afternoon, forty-eight hours later, the strike was for an increase of 5/- an hour on the basic and over 8,500 workers were out on strike. On Saturday and Sunday the local union officials and the shop stewards were urging the men to return to work: by the time of a branch meeting on Monday morning they had reversed their position and were declaring the strike 'official at branch level' (an empty formula according to union rules). For the first ten days Pilkingtons were expressing a willingness to negotiate but only if there was a return to work, and national officials of the General and Municipal Workers Union (GMWU) were publicly endorsing this position.

Dissatisfaction with the union had been expressed by some of its members right from the start and it grew apace during the first ten days of inactivity on the negotiating front. The strike spread to seven Pilkington factories outside St Helens embracing a total of 11,000 workers. The embryo of the Rank and File Strike Committee (RFSC), which was

INTRODUCTION

later to take control of the strike, was formed in this period and held its first meeting on Friday, 17th April – two weeks after the beginning of the strike.

By Monday, 20th April, Pilkingtons and the GMWU had agreed to an interim increase of £3 a week to be added to gross pay. This led to a return to work in the factories outside St Helens. But at a mass meeting of the strikers in St Helens on Tuesday, 21st April, the offer was rejected. It was put to another mass meeting three days later – and again rejected. After this the union ceased to be an effective source of leadership and the field was left to the RFSC. In the fifth, sixth, and seventh weeks the GMWU attempted to regain the initiative but without success. The dispute was not finally settled until Friday, 22nd May, when Mr Vic Feather, general secretary of the Trades Union Congress (TUC), stepped in with an offer of mediation.

The reaction of trade union activists on Merseyside to the Pilkington strike at St Helens, just 12 miles down the road from Liverpool, was one of disbelief. St Helens to them was in the Stone Age of trade unionism: St Helens people were known as 'furry backs' and had an unenviable reputation as strike-breakers. The strike has changed all that and much else as well. If there is one thing all parties to the dispute are agreed upon it is that: 'Things will never be the same again at Pilks.' And we might add that in so far as Pilks is St Helens, things will never be the same again in St Helens either. What happened in St Helens when the strike took place is the subject of this book. We think the story worth writing for a number of reasons, but principally because nobody has ever attempted a full-scale study of a strike in the United Kingdom.

Previous studies of strikes can be roughly divided into two groups.[1] Firstly, there have been those which have surveyed sections of the labour force in either one country or a group of countries. These have shown which parts of

1. For a more detailed summary of the strike literature, see J. E. T. Eldridge, *Industrial Disputes*, Routledge & Kegan Paul, London, 1968.

the labour force are most strike-prone and have then sought to identify the social, economic, and political conditions connected to a high rate of strike activity.[2] Secondly, there have been writers who have examined specific strikes or who have investigated particularly strike-prone firms or industries.[3] Usually these studies have attempted to reconstruct the events of the strikes in question after the disputes actually took place. Only one investigation, the justly famous *Wildcat Strike* by Alvin Gouldner, was actually carried out whilst a strike was in progress.

From these studies a number of themes have repeatedly emerged concerning the circumstances in which strikes are likely to break out.

Firstly, there is the interpretation of the strike as an emotional explosion of discontent inflamed by an accumulating list of grievances. Alvin Gouldner[4] found that the outbreak of his wildcat strike was preceded by the workers developing a feeling that management was treating them in an unfair and arbitrary manner. The firm involved had

---

2. Leading examples of this type of study are A. M. Ross and P. T. Hartman, *Changing Patterns of Industrial Conflict*, Wiley, New York, 1960; C. Kerr and A. Siegel, *The Inter-industry Propensity to Strike*, in A. Kornhauser *et al* (eds.), *Industrial Conflict*, McGraw-Hill, New York, 1954; K. G. J. C. Knowles, *Strikes*, Blackwell, Oxford, 1952; A. Rees, 'Industrial conflict and business fluctuations', in A. Kornhauser, *op cit*; E. J. Hobsbawm, 'Economic Fluctuations and some social movements', in *Labouring Men*, Weidenfeld & Nicolson, London, 1964; J. W. Kuhn, *Bargaining in Grievance Settlement*, Columbia University Press, New York, 1967.

3. The main examples of this type of study are A. W. Gouldner, *Wildcat Strike*, Routledge & Kegan Paul, London, 1955; L. Pope, *Millhands and Preachers*, Yale University Press, London, 1965; H. A. L. Turner *et al*, *Labour Relations in the Motor Industry*, George Allen & Unwin, London, 1968; W. L. Warner & J. O. Low, *The Social System of a Modern Factory*, Yale University Press, New Haven, Conn., 1946; P. Marsh, *The Anatomy of a Strike*, The Institute of Race Relations, London, 1967; Several analyses of particular disputes are also to be found in J. E. T. Eldridge, *op cit*.

4. A. W. Gouldner, *op cit*.

introduced new machinery and new managers. The combination of both meant new patterns of work and a changed relationship between management and men. The relationship changed over a two-year period from an easy-going style to a tightening up of rules, and punishment for their infringement. The result was a developing sense of grievance that eventually exploded in a strike.

From Liston Pope's [5] fascinating classic of the American South we learn how the changing market for textile products during the 1920's, the introduction of new machinery and the arbitrary reallocation of jobs, a cost-cutting economy drive, and a 'commodity' approach towards workers led to the growth of discontent and ultimately to industrial conflict.

A similar group of factors applied in the Yankee City shoe factory described by W. L. Warner and J. O. Low.[6] The background to this strike was a depressed market for shoes, the passing of ownership from local people to a big national corporation, and technical changes which entailed the replacement of hand skills by mass production methods. Clark Kerr and Abraham Siegel [7] suggested, after an analysis of strike activity in eleven countries, that strikes were most likely to occur in those industries which tended to monopolise a community: docks, mining, and seafaring. Since such communities are characterised by a similarity in type of work and level of skill, few opportunities occur for promotion within the industry and there are few chances of changing jobs. Furthermore, the fact that these tend to be isolated from the wider society engenders intense feelings of solidarity. Thus the social atmosphere generated by these conditions is such that unrest is relatively easily sparked off.

A second recurring theme in previous studies has been that the frequency of strikes is related to the efficiency of the machinery available for the settlement of grievances. If complaints can be laid and speedily resolved there are

5. L. Pope, *op cit*.
6. W. L. Warner and J. O. Low, *op cit*.
7. C. Kerr and A. Siegel, *op cit*.

likely to be fewer disputes. Thus Ross and Hartman,[8] on the basis of an inter-nation comparison of strike statistics covering fifteen countries in the period 1900–56, concluded that strikes are least likely to occur where the trade union movement is well-established, has a strong centralised organisation and a stable membership, where employers and unions have attained an acceptable balance of power, where there is a strong 'labour' party, and where governments provide conciliation machinery and take an active part in the management of the national economy.

From their study of *Labour Relations in the Motor Industry* H. A. L. Turner and his colleagues concluded that the high rate of strike activity was related to weaknesses in the organisation of trade unions and established procedures for negotiation and consultation.

A third type of strike-provoking situation has been identified by J. W. Kuhn[9] who argues that when a group of workers occupies a strategic position in the overall system of production it can, by withdrawing labour, use a very strong bargaining counter in its dealings with the firm. However, the circumstances in which a group of workers may be tempted to exploit its strategic position in this manner are not very common because pressure from other workers whose stability of employment would be constantly jeopardised is normally a restraining influence.

There are two sets of reasons why additional detailed studies of particular strikes such as the one that we launched during the Pilkington dispute are required. Firstly, the theories that have emerged from previous studies explaining how strikes are provoked need to be tested out by research conducted in the strike situation itself. It is only against evidence collected about the views and attitudes of people whilst they are actually participating in strikes that theories purporting to explain the incidence of strikes can be properly verified.

Secondly, case studies of particular strikes are needed

8. A. M. Ross and P. T. Hartman, *op cit*.
9. J. W. Kuhn, *op cit*.

because previous approaches to the study of industrial disputes have tended to neglect any analysis of the strike situation itself. Because they have been mainly concerned to present an explanation of strikes by referring to the characteristics of the firm and its market, the managers and its men, the strike itself has been pushed into the background. Previous studies have therefore tended to miss the point that a strike, while it is going on, is a power struggle. Now the way such a power struggle is conducted not only affects the course of the strike itself; it also affects the aftermath, and, even more important, is in some way or other a reflection upon the state of industrial relations prior to the dispute. Any strike amounts amongst other things to a crisis in an established system of authority. The Pilkington strike was a crisis not only for the firm but also for Britain's third largest union, for by the end of the stoppage the nucleus of a sizeable breakaway union had been created. Discovering exactly what happens during a dispute is clearly central to developing a full understanding of the significance of the strike.

We shall not make any exaggerated claims that our study enables all the outstanding problems in the field of industrial relations to be resolved. All we claim is that the type of case study that will be reported in the following pages can make a needed contribution to our knowledge of how and why strikes occur. We cannot claim that the dispute we chose to study was a 'typical' strike. Every strike is unique; no two of them ever occur in exactly the same circumstances.

In some ways the Pilkington strike was manifestly untypical. It lasted for the unusually long period of seven weeks, and it took place in a town whose labour market was dominated by the firm in which the strike broke out.

But whilst every strike possesses its own unique features there are enough common characteristics to merit some generalisations. It is possible to classify strikes into different types.

One obvious distinction is between official and unofficial disputes. The difference is usually cast in legalistic terms –

INTRODUCTION

a strike is official when it is organised and sanctioned by a trade union, and unofficial when it is neither organised nor sanctioned.

The difference, though, is more than one of rules. An official strike – unless it started unofficially and subsequently got union blessing – is usually the result of long-drawn-out negotiations which have eventually reached deadlock. In these cases the strike is strategically planned with defences well-dug and deployed in advance: it is akin to a military operation. There are, to be sure, some unofficial strikes planned in a similar manner, but there are many more which amount to spontaneous walk-outs or 'wildcats'. Clearly there is a big difference between an organised strike and a wildcat. In the first there is a declaration of war in the form of a strike notice. In the second there is a revolt in which all the normal constitutional channels are thrown to one side. In these circumstances it is by no means unusual for the demands to be formulated *after* the strike has started. It often appears as though the only way that strikers can make themselves understood is by putting a price on their return to work even though the original reason for coming out may have possessed only the most tenuous relationship to money.

The Pilkington dispute possessed all the distinctive characteristics of this type of unofficial, wildcat strike. Some of the unusual features about this stoppage resulted in the strike's 'typical' characteristics being thrust into distinct relief and therefore commended the strike as a particularly suitable subject for a case study. The fact that the stoppage was unusually prolonged meant that it was possible to conduct an effective on-the-spot investigation whilst the dispute was actually taking place. Also the length of time that the strike lasted resulted in the various parties involved and the roles that they were playing, becoming more clear-cut and visible than would normally be expected.

By examining how this strike originally broke out and how it matured into a prolonged struggle with issues and positions becoming increasingly clearly formulated, we feel that it is possible to offer some general suggestions about the factors

INTRODUCTION

that are involved in disputes of this sort.

Although this type of detailed case study of a strike is unusual the methods of research that we employed are not particularly controversial. The basic approach that we adopted in the investigation was simply to attempt to see the strike through the eyes of its different groups of participants. Translated into technical sociological language this means that we formulated our research within the framework of social action theory. By examining the strike from the points of view of each of its major groups of participants we have sought to understand why each group embarked upon particular courses of action, and thereby to explain the overall pattern into which the strike developed. As will become very clear in later chapters, the several parties quite genuinely and sincerely saw the strike in utterly different lights. What was real for one party was quite different from what was real for another. We shall show that this feature of the strike was fundamental in explaining why it occurred and why it was so prolonged.

Although we had discussed some months before the strike the possibility of doing such a study and had decided what our approach would be, it was obviously impossible for us to have planned this particular investigation and designed our research strategies in detail before the strike actually began. Techniques for conducting the fieldwork necessarily had to be devised rapidly in the context of the situation that presented itself. We will elaborate upon the methods that were used to collect our data as the results of the study are presented in the chapters that follow.

In outline, during the strike itself our research work consisted of a survey to discover the attitudes and opinions of a sample of 187 rank and file strikers that was conducted during the sixth week of the dispute, and observation of the activities of the RFSC during the latter half of the strike. In the weeks immediately following the strike lengthy interviews were conducted with all the members of the RFSC, with shop stewards and officials who were involved in the dispute on the Union's side, and with directors of the

INTRODUCTION

firm who had played central roles in formulating the company's policy throughout the dispute. One of us took up residence in St Helens for the last three weeks of the strike and spent the best part of fifteen hours a day either attending RFSC committee meetings, mass meetings, going on RFSC delegations, talking to people on street corners, in pubs, and on the picket lines. We made early contact with press reporters who very kindly declined to 'blow the gaff' on us and who provided us with all manner of snippets of interesting information – much of which regrettably we can make no use. We compiled extensive notes, and collected documents issued by the various parties during the course of the strike. All in all it was an extremely exciting, if tiring, business. If the strike was 'an education' for the people of St Helens, it was an education for us too.

This book has been written *by* professional social scientists but not just *for* other social scientists: we have tried to eliminate all jargon and write in such a way that it can be read by anyone who is interested in strikes in general and in this one in particular. On the other hand we naturally hope that our colleagues will not find it too 'slight'.

We believe it desirable that whenever possible social scientists should make themselves intelligible to those about whom they are writing. In our case we think we have a special responsibility because so many men and women in St Helens freely and willingly gave us large slices of their time and hospitality when many of them could ill-afford either.

Our decision to write in this particular way is also a 'political' one. In so far as books are written in specialist language they are only understandable to the initiated. The initiated tend to be those who have had a university or technical education, and those who have had this sort of education are usually in a better position to influence events than those who have not. The upshot of this is that those in influential positions have access to knowledge of potential political importance, while those on the receiving end of political decisions are deprived of this knowledge. In an era of growing political control over everyday life more people should find this worrying.

Throughout the book we have attempted to conceal the identities of our individual informants – they have only been named when they played such distinctive roles that it would have been impossible to conceal their identities. These people have of course given us their permission to do this. Otherwise we have avoided naming individuals partly in order to avoid unnecessary personal embarrassment, but mainly because we do not wish to personalise our account. We believe that the course of the dispute can best be explained in terms of the relationships that evolved between *groups* rather than in terms of what particular *individuals* did. The suppression of identities is therefore a realistic practice which aids rather than hinders the attainment of a full understanding. We have named the company, the union, and the town, for even if we had used fictitious names the dispute would have still been identifiable. We also felt that those reading our account should be able to judge for themselves the extent to which the particular background against which the strike took place entitles general conclusions to be based upon the study.

The book breaks down into three sections. In the first section we provide a general account of the setting and events of the strike. In Chapter One we outline the social, economic, and political background. We examine the history, the organisation, and the ideology of Pilkingtons and likewise of the GMWU. We look at St Helens: its development as the 'glass centre of the world', its labour market, the question of its being a 'company town', and we consider what sort of place it is to live in. Finally we examine the political and economic climate of Britain since 1945 and suggest ways in which this may have influenced the strike. It is *not* the purpose of this chapter merely to paint in a little bit of background or to add a touch of local colour. We consider that *any* social formation is only to be understood fully when it is related to the social, economic, and political structures of a particular time in history. This is a formidable

INTRODUCTION

enterprise but it must be treated as an essential part of the study. Chapter Two presents a strike diary, and its purpose is to provide an overall and concise account of the development of the strike, and to recapture the flavour of the public image of the strike that was being conveyed through the media. This diary concludes with a discussion of the role of the press during the dispute.

The next section of the book (Chapters Three to Seven) presents accounts of the strike as seen from the points of view of the rank and file strikers, the Union, Pilkingtons, and the RFSC respectively.

In the final section, Chapter Eight traces the immediate aftermath of the dispute, whilst Chapter Nine considers the general conclusions about strikes that can be drawn from the study. The Appendix that concludes the book contains reproductions of strike leaflets, correspondence, and other relevant documents.

One final word of introduction. Typically strikes are thought of as 'against the national interest', 'damaging to the economy', 'the work of a subversive minority', and so on. The label, whatever it is, is usually calculated to evoke the response of 'bad', 'wrong', or 'evil'. Since this is the usual view we want to make it absolutely clear at the start that we have no intention of producing a horror story. We shall not assess the justice either of this particular strike or of strikes in general. Nor shall we award victory to one side or the other. Our aim is simply to describe and to analyse. What we record here will be an inter-round summary rather than an evaluation of the dispute.

Part One

Chapter One

THE SOCIAL, ECONOMIC, AND POLITICAL BACKGROUND

St Helens

In 1750 St Helens was a chapel standing at the junction of four small townships – Windle, Eccleston, Parr, and Sutton (now suburbs of St Helens). In 1900 it was a large industrial town with a population not too far short of its present total of 101,000. In 1750 'lord of the manor' was a term that made sense; by 1900 'industrial baron' would have sounded more appropriate, for St Helens was nothing if not the creation of an industrial élite which grew to prominence on the coal lying beneath the ground on which St Helens now stands. Coal was king and was not to be deposed until after the Second World War.

The chemical industry, making alkali for the Liverpool soap industry, was established in St Helens in the 1820's; by the end of the century the industry was in a decline and by the 1930's had almost completely disappeared. Copper smelting was re-established in the 1830's but was in total decline by 1900. Plate glassmaking had been centred in St Helens since the 1770's, but flat window glass was not made on any scale until the 1830's. So far as the availability of raw materials was concerned St Helens was ideally placed for glass making – coal for the furnaces, and alkali and sand for the glass. It was also well placed for alkali manufacture – again coal for the furnaces, and salt was readily available from the Cheshire salt fields for decomposition into alkali and certain by-products.

St Helens was not, however, the product of the happy coincidence of certain raw materials. It also had people. Two sorts basically: industrialists and workers. The social history of nineteenth-century St Helens is on the one hand a history of the displacement of the old landowning class by a rising class of industrialists, and on the other the transformation of agricultural workers, migrant peasants, and artisans into an industrial working class. If a new ruling class was in the making in the shape of industrialists with their own characteristic sets of ideas, attitudes, and institutions, a working class was also in the making with *its* own ideas and institutions. The character of present-day St Helens is very much a product of these people.

The history of the industrialising families is an intriguing one. It is not just that they tended to marry each other's daughters, sisters, and cousins, though doubtless this influenced the way that families tended to dabble in a number of different industries. What is most fascinating was the tendency to have a finger in any pie that looked to have a good chance of making a profit. Thus the Greenall brewing family was, for example, variously associated with cotton spinning, banking, glassmaking, collieries, chemicals, landowning, property, and railways. The Pilkingtons, who were related through marriage to the Greenalls, were engaged, besides glassmaking, in distilling, chemicals, railways, and collieries.

While families usually only took a dominant interest in one industry, the firms set up in the first seventy years of the nineteenth century were typically partnerships involving only a few shareholders and fairly large amounts of capital. Many of the partnerships embraced at least one person with some technical competence, and these technicians were not often local men. Welshmen were associated with copper smelting, Scotsmen and French with glassmaking, Irish and Austrians with chemicals.

Several of these early industrialists, notably Peter Greenall and David Gamble (of the largest chemical firm), were also active in local affairs. The local landowners on the other hand, though still men of 'substance', were not remarkably so in this respect.

Peter Greenall had quite as much to do with the development of the town as he had with industry: he was prominent in St Helens Building Society (1824), the Gas Light

Company (1832), a waterworks company (1844), the Town Hall Company (1839), and he built St Thomas's church. He was elected Tory MP in 1841, was impressed with Edwin Chadwick's *Sanitary Condition of the Labouring Population*, and in 1844 had a Bill drafted to establish an Improvement Commission for St Helens which enabled the paving and draining of some streets together with the provision of gas and water. The Bill also stipulated minimum standards of housing, width of streets, powers of compulsory purchase, and would allow the new Commission to establish its own police force, to buy out the water, gas, and town hall companies, and to purchase the market-place from its Quaker owners. Opposition in St Helens succeeded in excluding some of the provisions (numbered amongst the opponents were the Pilkingtons) but the burden of the Bill became an Act of Parliament.

David Gamble came to some prominence at this time for he was associated with Greenall on the Improvement Commission and was later one of its chairmen. He became the first Mayor of St Helens when the charter of municipal incorporation was granted in 1868, and was already commanding officer of the local Volunteers: in 1861 he built their headquarters at his own expense. Gamble built the local technical college, The Gamble Institute, in 1896 (now the public library), stood as a Liberal in the 1885 general election against a member of the Pilkington family who stood as a Tory. Gamble lost. But the days of the open dominance of St Helens by its industrial and landowning élite were already numbered. A trades council nominee defeated a Pilkington in the local elections of 1890, and St Helens returned a Labour MP in 1906. Thereafter the industrialists were to discover that power could be exercised more subtly than through being politicians: to be a large employer was sufficient to ensure that the local council would be circumspect in its dealings. The personal need to live up to the Victorian ideal of 'service' was satisfied by sitting on the magistrate's bench.

Today's St Helens working class has, like its nineteenth-century ruling-class counterparts, very mixed origins. In 1800 its foundation was already laid: there had been colliers for several hundred years, and agricultural workers since time began or thereabouts. 'Foreigners' in the shape of Welshmen had already put in an appearance with copper smelting, and some local people had experienced the discipline of factory work in the plate glassworks established in the 1770's.

In the 1840's when there was a general shortage of skilled labour Pilkingtons brought in glassblowers from France. Other new arrivals at this time were the Irish: refugees from the famine. The settlement of the Irish had started in the 1830's and by the early '40's there were over 1,000 in the town. The Irish, like the colliers, lived in isolated communities – but for different reasons. The Irish experienced the traditional lot of the immigrant: isolation through ignorance and fear. (In one respect the Irish were fortunate in their 'choice' of St Helens for the area was a traditional Catholic stronghold – three of the four old lords of the manor were Catholics.) The colliers were isolated because they tended to live in cottages clustered around the pits where they worked, but as the small communities grew into each other with new house building the miners gradually ceased to be a 'race apart'. The integration of the Irish took much longer. Not until the end of the century, by which time they had started to inter-marry with local Catholics, were the Irish a settled part of the St Helens working class.

Given the divisions within the working class it is not surprising that it took so long to develop enduring institutions. It is true that a number of friendly societies providing sickness benefits and so on were well established by the 1840's. It is true also that there was a flurry of trade union activity in the 1830's – mainly among the miners – but there was little permanency in any of the labour movements until the last decades of the century.

A number of industry based co-operatives, not infrequently sponsored by the employers, made short-lived appearances in the '50's and '60's but the present-day St

Helens Industrial Co-operative was not started until 1883 by a group of bottlemakers: within sixteen years almost half of the population shopped at the Co-op. Trade unionism became firmly established in the same period – the Miners Federation of Great Britain was formed in 1889, there was a chemical workers union a year later, and several glass-workers' unions. Attempts to form a trades council had failed in 1867 and in 1885, but were successful in 1890: it had seventeen affiliated unions, four of them for glassmakers.

There were a number of notable strikes – most of them far outlasting the duration of contemporary strikes. The miners had two-month strikes in 1819, 1831, and 1844. The bricklayers and plasterers were out for twelve months in 1859–60, and Pilkington glass-blowers for six months in 1870. In 1886 one of the bottlemaking firms tried to reduce wages by bringing in Swedish workers. The bottlemakers struck and within two weeks the Swedes were on their way home – fares paid by the bottlemakers' union!

*

The rate of change in the twentieth century has been rather slower, although in terms of employment possibilities the change has been marked. It could not have been said in 1901 that St Helens was a company town, nor even in 1951 come to that. Not until the 1960's could that be said with any ring of truth when Pilkingtons became far and away the largest single employer.

In terms of population St Helens has entered into decline. In 1901 there were 84,410 people, in 1931 102,500, in 1951 110,000, and in 1966 101,000. In terms of employment probably the most significant change has been brought about by the motor car: the car has enabled people to travel further afield to work than ever before (though we should not underrate the degree of movement in more distant decades – no doubt the bicycle and cheap public transport earlier in the century were not without their effects). Thus in 1961 over 14,000 people living outside St Helens travelled daily into the town, while nearly 9,000 living in St Helens worked outside: a net inflow of 5,000 people. Over half of the people travelling into St Helens were manual workers. We do not know from how far afield these people came, but if Pilkingtons' estimate of their own labour force is any guide then people tend to come from within a twelve-mile radius. This area includes a large number of Liverpool overspill estates to the west and a number of old mining and textile areas to the north.

Over the last ten years the St Helens labour market has not undergone any major change – about 36% of the jobs have been in glassmaking (including white-collar employment). Mining had declined from providing 12% of the jobs in 1960 to about 6% of the jobs in 1969. The only other large employers outside the glass industry and mining has been Beechams, the pharmaceutical manufacturers, who employ about 1,000 people.

Most people work in factories employing more than 200 people: in 1969 there were fourteen factories and offices employing more than 500 people – seven of these belonging to Pilkingtons, and three of the others belonging to other glass firms – Rockware and United Glass the last remains of the old bottle-making and glass tableware firms. These two firms together account for the employment of about 3,000 people.

Is it possible to talk of St Helens as being a company town? Pilkingtons have not been great public benefactors: their factories and their offices are the only visible indications of their presence. The local council has been Labour dominated for over forty years, and if many of the councillors have been Pilkington employees, the great majority have been neither Pilkington employees nor members of the family. But, then, there has been no need for Pilkingtons to 'load' the council. No doubt Pilkingtons have given the council the odd nudge from time to time, but generally speaking the councillors and aldermen will not have needed reminding that Pilkingtons provide most of the jobs in St Helens and that anything affecting their interests required the most 'sympathetic and careful consideration'.

We would be very surprised if any of Pilkingtons' planning proposals had been rejected: the citizenry would have justifiably revolted had they been.

Pilkingtons' dominance is dominance of the labour market. It is not just that no other employer of male labour can afford to pay very much less than Pilkingtons, it is also a question of the peoples' feeling of dependence upon Pilkingtons. Nothing can be done to offend the firm too much, for who knows (?), they may pack their machines and take them somewhere else. In this situation people are likely to develop contradictory attitudes: an exaggerated deference on the one hand and a dull resentment on the other. Both attitudes were to reveal themselves during the strike which was why in some ways the strike took on a similarity to a peasants' revolt.

*

Outwardly St Helens is just like dozens of other northern industrial towns: closely packed terraced houses, lining gas-lit cobbled streets, penetrate right into the town centre; many of them without bathrooms or inside lavatories (in 1966 29% of the houses had no bathroom, and 34% had only an outside lavatory). Most people rent their houses: about 61% are rented as against 39% owner-occupied. The corporation is the biggest landlord, providing some 53% of the rented property. One of the largest 'private' landlords in town is the National Coal Board – whose houses are often occupied by people who have long since given up working in the pits.

The council housing estates are not much different from those to be found in any other part of Britain: those built in the 1920's and 1930's were laid out with reasonable economy of space and have been there long enough for the trees to grow. Those built since the war sprawl over treeless acres of unkempt green. But St Helens has its middle class, and it too lives in much the same way as the middle class in every other town, though its tidy and well-tended little enclaves are everywhere encroached upon by the march of

SOCIAL, ECONOMIC, AND POLITICAL

the corporation. The estates on the southern side of the town, the working-class side of the tracks, are punctuated by the debris of previous burrowings in the earth for sand, coal, and clay, and mounds of old alkali waste. Scattered around amongst this debris, on the odd level patch of ground, are the results of the corporation's attempts to attract small light industry: the odd clothing factory or two and a number of other plants, rarely employing more than a couple of hundred people, which make anything from electric light bulbs to earthenware pipes.

It is the town centre that defines St Helens. A grimy Victorian town hall without any of the fantastical flamboyance that distinguishes other Lancashire headquarters of municipal authority. A marbled and tiled library presented to the town some seventy years ago by a long-since-departed owner of a large chemical works. A Wesleyan church almost as large as the town hall and equally undistinguished. The town hall has the inevitable window-boxes of geraniums, the Wesleyan church a twenty-paces-long flower-bed precisely and methodically occupied by thin ranks of struggling French marigolds. A large Catholic church in begrimed stone and the usual Catholic confusion of styles is the town's only contribution to the grotesque. A couple of cinemas struggle for survival, while two others have given up and converted to bingo. The Theatre Royal, refurbished and subsidised at Pilkington expense, the only evidence of Pilkington benefaction, wobbles along on a mixed diet of Kenny Ball, Los Paraguayos, and an occasional farce provided by a visiting repertory company from Liverpool.

The only big shop in town, apart from Woolworths and Marks and Spencers, is owned by the St Helens Co-op; Boots is the only bookseller. Apart from the usual Tesco-type supermarket the shops are small and sell cheap goods: the covered market provides the 'bargains'. The busiest shoe-shop sells cheap shoes imported from Eastern Europe, and the menu of the busiest cafe ranges from plaice-chips-and-peas at 6/6d to beefburgers at 1/6d. Chippies abound and mostly thrive. The brewers thrive too, even if the landlords don't. In the town almost every street corner has its pub (and a betting shop not too far away): Victorian and Edwardian pubs. Central mahoganied bars with little parlours leading off: one of them 'reserved' for the women. Television in one bar is a concession to the solitary drinker and the elderly; a juke-box in another concedes to the oats-feeling young.

St Helens is a rugby league town, a man's town. The Women's Liberation Movement has not yet arrived: in St Helens the women do as they are told. They play bingo while the men go drinking. St Helens is a working-class town. The working class may not run it, but its culture is dominant.[1]

The glass empire of Pilkington Brothers

When the first glass was blown at the St Helens Crown Glass Company works in 1828, two Pilkington brothers were minority shareholders and the centre of British window-glass manufacture was on Tyneside. Within thirty-five years the Pilkingtons owned St Helens Crown Glass and no window glass was being made on Tyneside. Indeed by 1856 the flat-glass industry was dominated by three firms – Pilkington in St Helens, Chance in Birmingham, and Hartley in Sunderland: friendly relations prevailed between them. They fixed prices and bought up struggling competitors to close their factories and exclude new competition. Competition between British manufacturers was on the basis of quality and managerial skill: price competition came from foreign firms, mainly in Belgium. Belgian competition became particularly severe in the last quarter of the nine-

[1]. For the social and economic history of St Helens in the nineteenth century we have drawn freely upon T. C. Barker and J. R. Harris's excellent study, *A Merseyside Town in the Industrial Revolution St Helens 1750–1900*, Frank Cass, London, 1959. Other material has been taken from Census Reports. More recent information has been kindly supplied by the St Helens office of the Department of Employment and Productivity.

teenth century: Hartleys fell victim to the Belgians in the 1890's and Chances in the 1900's. Pilkingtons seemed to have survived partly because they had diversified into the manufacture of all forms of flat glass, particularly plate; partly because they had shown a constant readiness to adopt new techniques; partly because they had developed a large export market; and partly because the labour force was acquiescent.

Pilkingtons went through a lean period in the 1920's – they adopted a new method of making window glass inferior to that taken up by the Belgians. It meant the elimination of the skilled glass-blowers, but the glass could not compete with the Belgians for price and by 1930 Pilkingtons were considering giving up flat glass. They were saved by new techniques developed in the USA during the twenties – Fords developed a continuous process for making plate glass and a Pittsburgh glass firm invented a continuous process for making window glass. Pilkingtons adopted both – the latter in the early thirties – and by 1933 they were once again making window glass at a profit, helped at the same time by an imposition of an import duty on foreign glass. In the same year Pilkingtons bought out a firm making window glass at Queenborough in Kent, and Chances gave up making window glass altogether. In 1936 Pilkingtons acquired a large slice of Chances shares and by 1955 completely owned their old friends and rivals.

The period between the two world wars was for Pilkingtons one of international expansion and consolidation at home. New factories were built at Doncaster in 1922 and at Pontypool in the late thirties, shares were taken up in the new Triplex firm making safety glass for the motor industry, and a new range of products – glass pressings in the form of high voltage insulators – were first made at a new factory in St Helens in 1934. In 1938 the firm became the largest shareholders in a window-glass firm in Argentina, and in three safety-glass firms in Australia.

A similar pattern followed in the years after the Second World War. Fibre-glass, already produced by Chances in the 1930's, was being made in new factories in St Helens (1944), Glasgow (1948), and Birkenhead (1957). Optical glasses, another old-established province of Chances were being made at St Asaph in North Wales in 1957. And by 1958, after five years of costly experimentation, Pilkingtons had revolutionised the glass world by successfully making float glass. By 1965 Pilkington's policy of progressively buying up shares in Triplex Safety Glass bore fruit – Triplex became a Pilkington subsidiary when 51% of the stock passed into Pilkington hands. (Triplex had factories in St Helens, London, Glasgow, and Birmingham). In 1967 Pilkingtons acquired British Indestructo, Triplex's only major British competitor, and thereby became virtually the only supplier of safety glass to the British motor industry. This acquisition led to the Monopoly Commission Report on Pilkingtons in 1968.

International expansion continued in the post-war years. Window-glass factories were established in Canada and South Africa in 1951, and in India in 1954; controlling interests were taken up in safety glass firms in Brazil (1946), and New Zealand (1952). One loss was experienced: the St Helens colliery, owned since the 1840's, was nationalised in 1947.

By 1970 only two broad areas of glass manufacturing remained outside Pilkington control – bottlemaking and glass tableware, neither of which Pilkingtons had ever produced. No one was surprised therefore when Pilkingtons were the subject of a Monopoly Commission investigation in 1968. Few were surprised either when the report gave a favourable verdict: 'Our formal conclusions are, therefore, that neither the conditions we have found to prevail in relation to the supply of flat glass by reason of the monopoly position of the Pilkington group, nor anything done by the Pilkington group as a result of or for the purpose of preserving those conditions, operates or may be expected to operate against the public interest.'[2]

2. *The Monopolies Commission Report on the Supply of Flat Glass*, HMSO, 1968, para 307, p. 89.

SOCIAL, ECONOMIC, AND POLITICAL

But at least one person was surprised and said so – Alastair Sutherland, a Cambridge economist. He showed that the figures given to the Monopoly Commission by Pilkingtons understated by quite large amounts the true profitability of the firm. The understatement arose because Pilkingtons included the costs of developing float glass (£10½ million between 1953 and 1967) but neglected to set against this the revenue obtained from licensing the process to other manufacturers (estimated at £15 million between 1967 and 1971). Taking this into account Sutherland calculated the average rate of return on capital employed at 41% per annum for the five years 1961–65: markedly different from the 22% actually reported, and well above the average of Britain's 300 largest companies. Logically, this is what is to be expected from a firm that supplies approximately 90% of British flat glass, and virtually all of the safety glass consumed by the motor industry.

*

That Pilkingtons in 1970, one hundred and forty-two years after it first made glass, is still under family control – Lord [Harry] Pilkington is the fourth generation – is at least in part tribute to the family's ability to produce large numbers of children. Of the two original Pilkington glassmen, William, with his wife's assistance, produced six sons and six daughters, while Richard managed four sons. Of these, two sons of each took an active part in the business: these four sons together produced nine sons and fourteen daughters. From this third generation the firm captured eight of the sons, and three of the daughters made sufficiently good marriages to provide three sons suitable for fourth-generation directorships. The four other fourth-generation family directors are descended through the male line. The fourth generation who were board members in 1959 had produced seven sons and thirteen daughters: insufficient sons, when 'drop-outs' are accounted for, to ensure a family majority on the board in the fifth generation. A realisation of this may have had something to do with Pilkingtons' decision in 1970 to become a public company. However, 'outsiders' on the board will be nothing new for non-family men have been on the board since 1931: six of the board in 1970 are men who have 'made their way' in the company, albeit from managerial rather than shop-floor positions.

What sort of people have the Pilkingtons been? What has been their outlook on life and on the world? Above all, what sort of attitudes have they adopted towards their employees? We do not provide comprehensive answers to these questions for the Pilkingtons have been notably reticent and T. C. Barker, Pilkingtons' official historian,[3] hints at more than he ever actually says. Several things may however be said with a reasonable degree of confidence.

The Pilkingtons have been austere men, unostentatious. If they have lived comfortably, they have not lived flamboyantly: 'high living' has not been their style. Indeed, if anything, they have tended towards asceticism. There has obviously been a rare degree of dedication of purpose for the firm could not otherwise have grown to its present size and remained under family control. From reading T. C. Barker and other more recent statements by members of the family we are left with the very strong impression that the making of glass and the extension of the Pilkington empire has been endowed with high moral purpose, an attitude reminiscent of Nonconformist nineteenth-century capitalism. This characteristic in fact fits very well with the evidence we have. The first two Pilkingtons came from a dissenting family and although one of them turned Anglican, the other remained a Congregationalist. It is the Congregationalist side of the family that has provided most of the Pilkingtons active in the firm in the second, third, and fourth generations. The present chairman, Lord Harry, is an active Congregationalist and his brother, Lawrence, at one time worked with a missionary society. Lawrence Pilkington also wrote in 1934, remarkably for a man who is an extremely successful businessman, a long poem entitled: *The Dayspring: a*

3. T. C. Barker, *Pilkington Brothers and the Glass Industry*, London 1960.

protest against materialism. A number of living Pilkingtons are Anglican clergymen – three of them to be precise: one being Lord Harry's son, and another a second cousin. Lord Harry's second brother, Roger, is an ex-chairman of the Missionary Society.

Coupled with this moral dedication to the firm has been a severely paternal attitude towards employees and a sense of moral obligation to undertake public service in some form or other. Successive generations of Pilkingtons have been JP's on the St Helens bench, local councillors and aldermen, commanding officers of the local Territorial Army unit. Pilkingtons have not however been public benefactors – at least not in St Helens, although they did build a village to go with their new Doncaster factory in 1922. There have been no model villages, no art galleries, no museums, no parks, after the style of Cadburys or Levers.

Pilkingtons have attempted to exact from their employees the same devotion to duty that they have set for themselves. Those who have been faithfull have been rewarded – those who have been naughty have been smacked. (In 1878 a number of glassblowers and flatteners went on strike and a number of others stayed at work. In 1879, when contracts came to be renewed, those who had not joined the strike received more favourable terms than the others. In 1970 some strikers returned to work before the others. When redundancies were announced at Triplex after the strike the 'last in, first out' principle was waived for the strike-breakers. This action had union support.)

Pilkingtons, like most other British employers of the time, did not take kindly to the earlier trade unions. The American labour organisation, the Knights of Labour, sent representatives from the glassmaking centre of Pittsburgh to St Helens in 1884 in an attempt to organise Pilkington workers. They met with some success, but a Pilkington employee who chaired a meeting of glassworker trade unionists from Britain, France, Belgium, Italy, and the USA was subsequently sacked by Pilkingtons. It was later said – in 1887 – about the Knights of Labour organisation in

St Helens that they were in 'good organised condition' at other factories but that 'what with the tyrannical disposition of the capitalist and the general depressed condition of the men (through that curse Drink) we are not able to make any progress whatever... Messrs Pilkington Bros are continually discharging our members for no other reason than that they belong to our Society.[4] Other late nineteenth century attempts to establish trade unions may also have been stamped on. In 1889 the United Plate Glass Society started to enroll members in St Helens, but out of 800 members only 100 worked at Pilkingtons. A Sheet Glass Flatteners Society started in 1890 did not make much progress either. It was not until 1917 that any union was given recognition: to the National Amalgamated Union of Labour, and that union had been organising members in Pilkingtons for the previous twelve years. In 1964 an agreement was signed with the GMWU under which membership became a condition of employment.

From an early date Pilkingtons were making some sort of welfare provision for their employees although the extent of that provision at no time merited the exaggerated claims made for it. A sports and recreation club dates from the 1840's – the firm provided a field: a building was added in the 1860's. A schoolmaster was employed from the 1840's to teach the three R's to boy employees under seventeen – but the boys attended in their own time and were fined for absence! The fines were used to buy books for the works Library. A cycling club was formed in 1891 – the firm bought bicycles at cost and sold them to members by weekly instalments. Works canteens date from the 1880's, as does the medical service which, interestingly, was tacked on to a friendly society already organised by the glassblowers. The firm asked the blowers to extend membership to all employees and the firm in return started to subsidise the fund. In the period between the two world wars dental, optical, and chiropodists' surgeries were equipped and rented

---

4. T. C. Barker, *op. cit*, p. 178.

out to local practitioners – so that employees would not lose too much time off work when attending for treatment. At the same time a fully equipped operating theatre was provided as were physiotherapy and rehabilitation services – these provisions being related to the fact that glassmaking can be a very dangerous job. Apart from the sickness, injury, and superannuation schemes the only further welfare provision of any note is the assistance now given to pensioners – a delivery of free coal at Christmas and a contribution towards the cost of a television licence. One great object of pride to Pilkingtons has been its Joint Industrial Council – a consultative body consisting in equal numbers of elected shop-floor representatives on the one side, and appointed managers on the other. This dates from 1918, and was set up under the instructions of the then Ministry of Reconstruction. Government policy changed subsequently but Pilkingtons kept the Council in being.

In 1955 H. A. Cozens-Hardy, a managing director of Pilkingtons, said: 'The firm's underlying policy toward its employees is ... one of mutual understanding and trust.' The cost of implementing the policy was at that time about £75 per annum per head of which £42 was the prosperity bonus paid to *staff* employees and the cost of holidays with pay. Thus of the remaining £33, if it were taken into account that the term 'employee' included managers and office workers who were in receipt of far superior fringe benefits than the manual workers, the actual extent of munificence extended to manual workers was not particularly remarkable and certainly no better than that provided by many other large British firms.

*

The Pilkington conception of their firm was very well expressed by H. A. Cozens-Hardy when he lectured to the Royal Society of Arts in 1955 on *Welfare in a Large Family Firm*: 'The firm is, in fact, a family firm in more senses than one. It is a firm which for over a century and a quarter has been owned, directed, and managed by the Pilkington family, and it is a firm in which the family feeling is also experienced by the employees themselves. A large proportion of the population of St Helens is dependent upon the prosperity of the firm, and there are many families who have worked for two, three, or even four generations. The fact that there are so many families engaged at all levels, gives the integrated structure which is of great value in preserving the family spirit throughout the Pilkington group and may well give a settled pattern of affairs which makes for harmony. It has certainly resulted in a mutual regard which has encouraged the Pilkington family to devote so much attention to the welfare of employees.'[5] But earlier on in the same lecture we find that a different reason is given for providing welfare benefits: 'It is rapidly becoming appreciated that although machines, equipment, power, and materials may be regarded as units of production, the conception of units of manpower cannot be accepted, for the efficiency of those at work depends on their temperament, their surroundings, the action of their fellow workers, and the attitude adopted towards them by those in authority.'[6]

This means that workers *are* as a matter of fact units of manpower but they must not be *treated* as if they are. This of course is sound policy because workers do not generally like to think of themselves as commodities even though the accountant doing his sums on the computer prices labour in exactly the same way as he prices raw materials and depreciation of plant, etc. Translated a bit further: treat your workers as if they are people and then perhaps they will not notice that they are commodities. The talk then of the 'family spirit' is a fine piece of self-deception – but none the less sincerely believed for all that.

The 'severe' tone of the Pilkington style is exemplified in the harsh and peremptory tones of the *Time Recording Regulations* (reproduced in the appendix) and in the emphasis placed on discipline and 'responsibility' in several of Lord

---

5. *Journal of the Royal Society of Arts*, 1955, p 362.
6. *Op cit*, p 353.

Pilkington's annual reviews for employees.

In 1967 he is saying: 'I remain convinced ... that if we are going to balance our payments at a high level and look forward to a thriving future, then we in this country and in some other Western countries must not merely work harder and better to improve our general productivity, but we shall also need to work more. Whether this be through longer hours and less idle time in the day, or rather more Saturdays or rather fewer holidays in the year, or even a later age of retirement and pension is a matter of opinion.' In 1968 the trade unions are being thanked for their assistance: 'I believe that they know when employers really want to work with them to the betterment of all ... and that when that is the case trade unions will play their part, not merely in pressing for more regardless of conditions in the outside world, but in helping to ensure the discipline which makes it that much easier to advance social progress.'

By 1969 the message is both sombre and prophetic: 'Glassmaking is a matter of discipline ... the more people are linked to one another through manufacturing processes the more important is that discipline ... it is our aim to ensure discipline where it matters, and to give freedom where the interests of others are not endangered.' Later in the report the same theme returns: 'If management has to be disciplined in its approach to wage negotiations and settlements, we must also do our utmost to support trade unions in their crises for they too are suffering from a falling away of effective discipline ... In negotiations there must be no doubt of the authority of those who negotiate or of their ability to carry out to the full their side of any bargain. We have consistently supported unions when it has not been too difficult to do so, but now when times are more difficult it is even more important to help the unions to maintain or restore discipline in their own ranks.' And then in discussing the proposed productivity programme: '... the productivity programme (is not) the answer to everybody's problems. It could, for instance, have little impact in the next twelve months – the most critical time. Therefore, during that time it must be by personal sensitivity, and by watchfulness and initiative that we avoid unecessary troubles through allowing quite small local issues to lead to great big trouble spots for us, constructed by those who are really ill-disposed – and we must not forget that there are some such about.'

Clearly Lord Pilkington has been disturbed by strikes in the motor industry, but as regards his own company he is really more optimistic: 'No company starts with more goodwill than we do ... let no one ever say that we are arbitrary or authoritarian ... We are already nearer, much nearer, to a new system than others ... In fact our greatest asset is the century of personal contact and established goodwill of which I receive evidence very often – wherever I go. There is nothing to be ashamed of, much to be proud of, and therefore we can pioneer in partnership as so often before.'

In these short quotations we have the very essence of the Pilkington style: the emphasis on hard work, discipline, responsibility, and partnership. What is significant is perhaps not so much that there is a continuing restatement of the necessity of these qualities (after all, few would disagree in principle that at least some of these qualities *are* necessary in some degree for the running of an organisation), as that Lord Pilkington should feel secure enough in his position to be so blunt and forthright. This forthrightness, we would suggest, is a function of two things. Firstly, a conviction of moral rectitude uncomplicated by doubt. And secondly, and more importantly, because the prevailing system of authority has never, in the history of the firm, been seriously challenged by its employees. Only a few moments' thought on the implications of these two factors says much of what there is to say about the background of the strike.[7]

---

7. For this reason we have drawn heavily on T. C. Barker's *Pilkington Brothers and the Glass Industry*, which, although inevitably suffering from the defects of being an 'official history', we nevertheless recommend as the only reliable source on Pilkingtons. See also: A. Sutherland, *The Monopolies Commission in Action*, Cambridge University Department of Applied Economics Occasional Papers, No. 21, 1968

SOCIAL, ECONOMIC, AND POLITICAL

Life and labour on the shop-floor

Five of the six plants in St Helens make glass products of one sort or another; the sixth, Triplex (1,200 employees approx.), toughens glass by a reheating process mainly for the motor industry. The Ravenhead works makes a variety of glass pressings – cathode-ray tubes for TV sets, high-voltage insulators, decorative glass blocks, etc. (1,000 employees approx.). The Fibreglass plant (900 employees approx.) produces, amongst other things, fibreglass mouldings, fibreglass matting, and fibreglass filament. The Sheet Works (2,400 employees approx.) makes window glass of different types, e.g. plain glass, opaque glass, wired glass, figured glass, etc. Cowley Hill (2,000 employees approx.) is the home of the float process, a method of producing plate glass by floating a ribbon of molten glass on a bed of molten tin. A fair proportion of this glass goes to Triplex for toughening. City Road (500 employees approx.) silvers glass for mirrors and bevels it, as well as producing double-glazing units. At the time of the strike the firm employed some 8,200 process workers, i.e. those actually engaged in making or toughening glass. Of the remaining 7,000-odd Pilkington employees in St Helens, 1,600 or so were maintenance and transport workers unaffected by the strike.

The making of glass like most other basic materials is, in principle, quite easy. To oversimplify more than a little: mix sand, lime, and soda, stir well and melt to a temperature of about 1,600 °C. Cool slowly and remove from oven. The only complication is that to produce it cheaply, in large quantities and of consistent quality, very large amounts of capital are required, and the process must be non-stop (168 hours a week). Because the process is continuous shift work is inevitable, and because it is capital intensive (i.e. small amounts of labour mixed with large amounts of capital equipment) there are very few jobs that require high levels of skill: at least 90% of the strikers were semi- and unskilled workers. Some of the work is heavy, some involves great heat, some is dirty, and some is dangerous.

LIFE AND LABOUR ON THE SHOP-FLOOR

The strike started amongst a group of glass carriers in the Flat Drawn department of the Sheet Works where work can be both dangerous and hot – it is sometimes called the 'blood-shop'. Flat drawn makes window glass: a hot but cooling ribbon of glass is drawn vertically from the furnace (tank) through a three-storey building, cooling as it rises. On the top floor the ribbon is machine cut, and the glass lifted away by men using vacuum suckers. These men are the glass carriers.

Hot air rises from the tanks so that, especially in summer, glass carrying can be extremely uncomfortable: the men's clothing may be literally saturated in sweat from head to toe. Like merchant seamen trading to hot climates, the glass carriers and others on hot work are supplied with orange or lime juice and salt tablets. Glass carrying can also be dangerous – if the glass has not cooled to the right temperature, or contains an imperfection, by the time it reaches the top floor it may shatter as soon as it is lifted off. There have been some pretty gruesome cuts amongst the glass carriers. There has also been a sizeable turnover of labour – of the forty-five men who set the strike in motion thirty-three had worked for the firm for less than a year and twenty-four for less than six months.

Working conditions do of course vary enormously: if there are workers in Fibreglass who suffer from skin irritation ('You can recognise a Fibreglass worker by the fact that he is always scratching'), and others in Ravenhead who get blackened by fumes from burning oil brushed over the plungers in the glass-pressing machines, there are thousands of others whose jobs are neither remarkably unpleasant nor physically dangerous. With respect to working conditions it should be pointed out that the Pilkington research establishment devotes some of its energies to developing protective clothing, and that the Ergonomics Department is perpetually engaged in trying to improve conditions.

As the firm has expanded, its organisational structure has undergone profound changes, not least with regard to the handling of industrial relations. Prior to the reorganisation on a product division basis, i.e. each division being responsible for the production and marketing of a distinct range of products such as flat glass, fibreglass, etc., industrial relations tended to be handled centrally and more or less single-handedly by one man who was a board member. Since reorganisation industrial relations has continued to be centrally controlled, but by a specialist department – Group Industrial Relations – which is responsible to a board member. At works level the works manager, in consultation with personnel officers, is responsible for all industrial relations questions: he cannot however negotiate in such a way that agreement would entail cutting across prior JIC agreements. In day to day operations Group Industrial Relations does not appear on the scene unless a shop steward has called in a full-time union official to help him in his approaches to works management.[8]

During the strike the blame for the eruption came to be laid by all parties, including the Court of Inquiry, at the door of the wages and negotiating structures. (Some senior executives were sceptical about this – they pointed out that the wages structures of other firms known to them were equally complex.) Negotiating procedures were not so much complex as highly centralised in the Joint Industrial Council (JIC). The JIC had 44 members – 22 appointed by management, 15 by GMWU shop-floor workers from all the GMWU organised Pilkington plants throughout Britain, and 7 full-time officials appointed by the union. All members sat for periods of two years, the shop-floor representatives being elected by the shop-floor. Either side of the JIC had powers to raise almost anything they liked. If the main concern was with wages and bonus levels, the JIC's constitution enabled discussion of anything else from encouraging research into manufacturing processes, to publishing '... authoritative statements of public interest upon matters affecting the Industy'. The constitution, in short, was so drawn up as to allow the discussion of most of what went on

8. These statements apply to the situation that existed before the strike.

inside the firm. Whether or not precedents had been made which effectively precluded certain matters is something we do not know.

As far as wage questions were concerned the JIC was a cumbersome body which, because it was concerned to ensure uniformity throughout all the British plants, made local, plant-based negotiations difficult. Shop-floor workers could initiate JIC negotiations through their representatives. Shop stewards could also exert some influence upon job evaluation studies which would affect earnings, but plant bargaining as such was beyond their powers.

Over 80% of the process workers were on two different systems of payment – direct system workers were on a piece-work system, i.e. a base rate of pay plus a bonus related to the actual output of each individual worker, the bonus being a percentage of basic hourly rates. The other large group of workers – about 45% – were on a multi-factor bonus scheme. Workers on this system, because of machine-controlled processes, were less able to control output. Bonuses therefore were related to such items as manning strengths, output, and speed of process. Workers on this system were, by and large, less well-paid than those on the piece-work system. In addition to these bonus systems all eligible workers were paid, on top of base rates, increments for shift-working and overtime. Wages were based on a $37\frac{1}{2}$- or 40-hour week for shift-workers, with three different shift systems being operated. Two of the shift systems guaranteed a free weekend every week (apart from overtime working), one system would guarantee 48 hours off every week, but only one 48-hour period in any one month would fall on a weekend. Shift times were: 6 am to 2 pm; 2 pm to 10 pm; and 10 pm to 6 am. Apart from a minority of day-workers no one would be working the same hours for two weeks running. No one either could be certain that free-time was in fact free-time for 'the group expects you to work overtime as required'.

Average earnings for males over twenty-one years was the best part of £29 a week for 46·8 hours. In October 1969 the process workers were paid about £26 10s for 46·8 hours on average – workers in the vehicle industry (including maintenance workers) were receiving on average about £29 for 43·6 hours, while workers in paper, printing, and publishing were getting about £29 for 46·1 hours. These were the only two *industries* paying more than Pilkingtons.

Averages can of course be misleading as the following figures suggest (figures taken from the Pilkington submission to the Court of Inquiry):

Gross pay	40–44 hours	Hours worked 45–49 hours	50–54 hours
£15–20	182 men	138 men	nil
£20–25	684 men	229 men	170 men
£25–30	468 men	434 men	157 men
£30–35	277 men	456 men	166 men
£35–40	67 men	155 men	59 men
£40–45	12 men	54 men	40 men
£45–50	nil	17 men	12 men
Total	1,690 men	1,483 men	604 men

Excluded from the above table are the 29% of process workers in St Helens who worked less than forty hours, and the 16% who worked more than fifty-four hours per week. Even amongst those who worked around the average number of hours (46·8 hours) there were many who lagged behind the average level of earnings. The overall average was pulled up by minorities of men who either worked exceptionally long hours or who were exceptionally well-paid.

The wide spread of wage levels was due partly to the operation of different wage payment systems. The disparity between the payment of the workers on piece-work and those on the multi-factor bonus scheme was well illustrated by Pilkingtons' estimates of the effect of a 6d an hour increase on basic rates. A multi-factor bonus worker on three shifts continuous working would have got, without overtime, an increase of 29/2d per week. A worker on piece-work, working the same shifts and on the middle bonus range, a 36/8d a week rise – 24% more. This does not mean that all piece-workers were 24% better paid than other process workers, but it does suggest disparities of wide margins in many cases.

The General and Municipal Workers Union: an empire of labour

The National Union of General and Municipal Workers, to give it its full title, was formed in 1923 out of an amalgamation of three unions: the National Amalgamated Union of Labour dating from the late 1880's, the National Union of Gasworkers and General Workers established in 1889, and the Municipal Employees Association started in 1899. The new union, the GMWU, had a total membership of 315,000 of which the Gasworkers had contributed 221,000, the National Amalgamated 53,000, and the Municipal Employees 41,000. The constituent unions had different traditions. The National Amalgamated and the Municipal Employees being essentially conservative by adopting conciliatory attitudes towards employers: the early leaders of the National Amalgamated were Liberals. The Gasworkers on the other hand had been an aggressive and militant union. Its first general secretary, Will Thorne, was a militant socialist and a prominent member of the Social Democratic Federation, the first British party to be strongly influenced by Marxist ideas. Thorne was a friend of Marx's daughter, Eleanor, and there is reason to believe that she gave Thorne a good deal of help in the formative years of the Gasworkers Union. (It was by no means unusual in those days for middle-class socialists to lend their skills to the new and struggling unions of semi- and unskilled workers.)

The period between the wars was a lean one for trade unionism. Depression and widespread unemployment (18% of the St Helens working population was unemployed in 1931) made aggressive trade union activity practically impossible to pursue. The most the unions could do was to resist wage reductions and hang on to the membership they had. The GMWU lost members just the same – as did every other union. Between 1924 and 1933 membership fell by 33% from 360,000 to 241,000. By 1937 the worst of the depression was behind and membership started to increase again – 420,000 in 1937. With a return to nearly full employment during the Second World War and the maintenance of full employment in the post-war years the unions entered the period of their greatest expansion: GMWU membership increased by 73% between 1939 and 1951 to 809,000 members. Membership declined thereafter, but not appreciably: 798,000 in 1969. In the same period the Transport and General Workers Union, the other big general union, increased its membership from 1,285,000 in 1951 to 1,475,556 in 1969.

The structure of today's GMWU does not differ greatly from that of 1923. All members of the union are attached to a branch which is either based on their place of work or the area in which they live. (In 1965 there were 2,261 branches of an average size of 350 members. Sixty-six branches had 2,000 or more members and these accounted for 25% of the total membership.) The branches come under the jurisdiction of a regional council whose members are elected from the branches. There are ten regions and these are administered on a day-to-day basis by a full-time official in each. There were altogether in 1965 a further 140 full-time officials spread over the ten regions.

At national level there is an annual delegates congress which lays down the broad areas of policy – delegates are elected from the branches and the regions on the basis of one for every 2,000 members of a region and one delegate per branch. Full-time officials may attend the congress and speak but not vote. Between congresses the union is administered by a General Council consisting of the chairman of the union (elected by congress every two years), the general secretary (elected in the first instance by a ballot of all members and then retaining the position for life), the ten regional secretaries, and fourteen lay members (elected for two years from the regional councils). The General Council meets quarterly but appoints a National Executive Committee to transact any of its business and to make decisions on its behalf. The National Executive Committee (NEC) is composed of one representative from each region, half of whom must be lay members. It meets monthly and in 1965 had seven sub-committees.

There has only been one major constitutional change since 1923 and that was with regard to the election of full-time officials. Until 1926 all full-time officials had to submit themselves for re-election every two years. After 1926 all officials, once elected, were elected for life. Officials were to be *appointed* in the first instance for a two-year probationary period at the end of which they had to submit themselves for election. This system still obtains. The branches may nominate lay members to contest elections, but the NEC decides whether or not they shall be allowed to contest an election, for Rule 19 states: 'No member shall be eligible for nomination and election to a Regional position unless his or her qualifications are in keeping with the standard required by the Union (such standards to be determined by the National Executive Committee in its absolute discretion) . . .'

The general secretary is responsible for the administration of the union and under him are eight national industrial officers (appointed by the NEC from the regional officers) and a national women's officer. In 1965 the head office staff amounted to ninety-seven employees distributed over six departments. In 1965 the research department had a staff of eight, six of whom were graduates. The educational department spent £56,000 in 1965, most of it on the union's residential training college for officials and shop stewards. The bulk of the remainder went on scholarships for members attending Ruskin College and on bursaries for the sons and daughters of members attending university.

*

In giving evidence to the Donovan Commission on Trade

Unions, the GMWU claimed to be democratic – a claim since contested by the Pilkington strikers. If democracy means rule by the people then clearly the GMWU and every other British trade union are undemocratic for patently the union members do not rule their unions. But most people's conception of democracy seems to mean the election of people to represent them, the right to remove their representatives if they prove inadequate, and the toleration of opposition. Even on this conception of democracy it is clear that the GMWU, in company with most other unions, is not democratic. The officials once elected, stay elected, and can only be dismissed by the National Executive Committee, itself half composed of 'elected' officials. (There have been instances of dismissal – Arthur Lewis, at present a Labour MP, was sacked in 1948 after his conduct during the famous Savoy Hotel strike in 1946: he led an unofficial strike, although the strike was subsequently made official.) The union is also in a position to determine who shall be elected, for the NEC is empowered to determine the 'suitability' of candidates. (See Rule 19 quoted above.) Opposition within the union is not tolerated – under Rule 43 any member '... who makes or in any way associates himself or herself with any defamatory, scurrilous or abusive attacks, whether in any journal, magazine or pamphlet, or by word of mouth, on any official of the union or Committee of the union, or who acts singly or in conjunction with any other member or persons in opposition to the policy of the union as declared by its Committees or Officials under these rules, or for any other reason deemed good and sufficient ...' is liable to expulsion.

The GMWU did not explain to the Donovan Commission how it justified its claim to be democratic, but then the members of the Commission did not seem to be at all interested in whether or not trade unions were democratic: it is a subject that was never once mentioned in the final report. Presumably the GMWU would justify its claim by referring to the fact that union policy is laid down by a delegates conference of lay members. If such a claim were to be made it would be difficult to substantiate, for it implies a neat distinction between *making* policy and *administering* policy. This distinction, as every good civil servant knows, is mythical for the implementation of a policy considerably affects its operation. In the GMWU as in other unions, policy is implemented by full-time officials over whom lay members have virtually no control: if the lay members lay down policy, it is the officials who decide how that policy shall work. Some unions, notably the Transport and General Workers, have sought to place more power in the hands of the lay union members without making any drastic constitutional changes. They have done this by the simple expedient of referring agreements made with employers back to the shop-floor for their approval. This policy has not met with the approval of GMWU general secretary, Lord Cooper. In February 1970 when the Transport and General Workers negotiated an agreement with Fords they referred it back to the shop-floor for approval: Lord Cooper regarded this as an abdication of responsibility.

*

A lot was made of the GMWU being a family concern just like Pilkingtons during the strike in St Helens. There is some truth in this: Lord Cooper's uncle was Lord Dukes, a previous general secretary of the GMWU. David Basnett, a GMWU National Industrial Officer, has worked for the union ever since he left school: his father was a full-time official. Jack Eccles, another National Industrial Officer, is third generation – both his father and his grandfather were full-time officials of the union. Fred Hayday, yet another National Officer, has never known any employer other than the GMWU: his father was also a full-time official. And Lord Williamson, GMWU general secretary from 1946 to 1961, had an uncle who was a full-time official. There has clearly been a considerable degree of nepotism – dating from the time when the local officials would give the clerk's job to a son – but too much could be made of these family connections. What is more important, as with Pilkingtons, is to present an outline of the GMWU attitude toward industrial relations and its members.

In two major respects the GMWU closely resembles Pilkingtons: in its emphasis on partnership and on discipline. When giving evidence to the Donovan Commission Lord Cooper made two highly significant remarks. Firstly: 'It is an elementary requirement of our basic purpose that we should do everything possible to contribute towards maximising the revenue of a firm or industry to increase the prospects of obtaining better wages and conditions. This approach is the basis of the fruitful co-operation which we enjoy in many firms in which we have exclusive, or near exclusive organisation of manual workers.' And secondly: 'We consider that industrial relations would be significantly improved if more firms regarded trade unions and collective bargaining as valuable instruments in promoting the objectives of the firm to everybody's benefit.'[9]

This indicates an attitude of mind that would have been unthinkable to the men who started the Gas-workers Union for it is to affirm what they used to deny: a harmony of interest between employer and employee. It is also in striking contrast to the views expressed in the new Transport and General Workers Union shop-stewards handbook, where it is stated that the interests of workers and managers are not identical.

On matters of discipline the GMWU has quite as strong a line as Pilkingtons: it is unalterably opposed to unofficial strikes and equally firm about its obligations to punish those who offend in this respect. In at least one case the GMWU has signed an agreement with an employer to the effect that unofficial strikers would be sacked with the union's blessing. This agreement was with Ilfords, the photographic suppliers, in 1965. Ilfords agreed to a GMWU closed-shop, and in return the GMWU undertook to get its members to sign individual statements unreservedly agreeing to abide by union rules and agreements by the union and the com-

9. *Donovan Commission: Minutes of Evidence*, p 1780.

pany. At the time a GMWU official said: 'A greater degree of self-control is needed and members must be taught to honour agreements.'[10] As the *Financial Times* said: 'The statement which process workers are being asked to sign makes it clear that, should any undertaking be broken by individual members, Ilford can terminate the agreement. Thus, if there is an unofficial strike and the GMWU does not discipline its members involved it could lose its right to 100% organisation among process workers.' Ilford's chief personnel officer, when asked what the men got out of the deal, said: '... the union has been pressing for years for 100% membership, and the union represents the men. So the men get what they asked for.'

What emerges from all this is that there is a distinctly professional aroma about the GMWU: that is to say effectively it regards its members as *clients* rather than as *participants*. The 'members' pay their dues and in return are provided with certain services – provided they do as they are told. A very distinct line is drawn between the union and its members – the union becomes the organisation, its corps of full-time officials. The members become passive dues-payers who may be allowed a vote now and again (but they can only vote for people approved of by the 'union') and to send delegates to a conference which effectively has very little power. The union is in other words very much in the charge of its officials, officials over whom the lay membership has little control since they are not subject to periodic re-election.

The GMWU seems to have been very sensitive to the criticisms of middle-class intellectuals of the trade union movement as a whole and to have adopted many of their proposals. The GMWU employs more people with an academic training than any other union, it has taken up large blocks of shares in public companies – Cooper and Hayday are directors of Yorkshire Television, it has a high subscription rate, and provides extensive training facilities

10. See *Financial Times*, 29th October, 1965, for a full report.

for shop stewards and officials. Unfortunately for the lay members, the intellectuals' criticism (of which the Donovan Report provides an excellent example) has been based on the assumption that the unions have been too weak and ought to be strengthened. But the 'unions' have been narrowly defined in terms of the union *apparatus*, that is its administrative apparatus, and the consequences of this narrow definition have been recommendations aimed at increasing the power of the officials over the membership. Thus, while the intellectuals' prescriptions have been aimed at creating an industrial environment freer from conflict, they have actually succeeded in prescribing strife. By concentrating on union administration they have ignored the fact that there are still sufficient rank and file trade unionists about who think that unions ought to be democratic and in the last resort under the control of the members. A revolt on the part of the rank and file has been on the cards for some time: that it should have led in the case of the GMWU members at Pilkingtons to the setting up of a breakaway union is not then surprising, for the GMWU, by taking up the intellectuals' prescriptions, has been sowing the seeds of revolt for at least a decade. We shall be returning to this theme in the concluding section of this chapter.[11]

The trade unions and the power structure

In one major respect the policies of successive British governments since the war have been identical – the overriding concern of all of them has been to maintain full employment. Governments have not always found this easy for it has involved from time to time such unpopular measures as tax increases, bank rate increases, and 'pay pauses' or 'wage freezes'. In the main governments have

11. For this section we have relied largely upon H. A. Clegg *General Union in a Changing Society*, the GMWU Rule Book as revised in 1969, and the GMWU evidence to the *Royal Commission on Trade Unions and Employers Associations* (the Donovan Report) *Minutes of Evidence No 42* London HMSO 1967.

been successful although at a price of continuing inflation and a low rate of economic growth.

By the late 1950's economic growthmanship was becoming fashionable amongst economists and politicians, a fashion not unconnected with the fact that the periodic deflationary measures required to curb inflation also hindered growth. It became obvious that the twin objectives of growth and curbed inflation could only be attained if increases in prices and incomes were related to increases in output. The attainment of these ideals required an over-seeing role on the economy of some magnitude. It also required the co-operation of the two main economic parties – business and trade unions. The first attempt to secure this co-operation on any scale came in 1962 with the establishment of the National Economic Development Council (NEDC). The TUC, after playing 'hard-to-get' for four months, decided to join in.

By joining this nascent planning body the TUC was not departing from any tradition of co-operation with either business or government. Informal contacts between trade union leaders and businessmen had been well-established for many years and they had frequently met each other on committees at both a national and local level. Trade union leaders were no strangers to government either. Although the trade unionists were being drawn into the corridors of power as early as the First World War the process was never as complete as it has been since 1940. Since then TUC nominees have sat on numerous advisory committees attached to various government ministries, and have developed strong informal links with the Ministry of Labour (now the Department of Employment and Productivity), and the Treasury.

While the debate over whether or not the TUC should join the NEDC was going on two strands of trade union philosophy emerged. One school of thought had it that the NEDC's job was to prop up a tottering capitalist system and that trade unionists should therefore have nothing to do with it. The other school was no longer thinking in terms of the desirability of capitalism and of conflicts between labour and capital, but of the 'national interest' – as Lord Citrine, a former TUC general secretary, said in the House of Lords on 8th November, 1961: 'The council could work not merely in the interests of the nation but in the fundamental interests of trade unionists themselves.' In the same speech Citrine (at one time a Pilkington worker) indicated his feeling of hurt that the government had failed to consult the TUC before implementing its pay pause of that summer.

Those trade unionists who were urging non-participation on the grounds that it involved consorting with the enemy had missed the boat by many decades for Winston Churchill had well summed up the position of the trade union movement many years previously when he had said that the trade unions were now the fifth estate of the realm. Meaning of course that the trade unions were integrated into the ruling apparatus of Britain. This is truer now than ever: trade unionists are to be found wherever power is exercised. On the NEDC, on the Prices and Incomes Board, on the Industrial Reorganisation Corporation. At a more local level they sit on the NEDC's associated with particular industries, on regional economic planning councils, and on Courts of Inquiry into strikes. In fact wherever there is a public body, extending from the governing boards of nationalised industries to the local magistrates bench, there is a trade union official sitting as a member. There can be no argument about it: the trade union administrators have arrived.

While business has had the ear of government for as long as there has been business, it too has been brought into ever closer contact – an almost inevitable development in a managed capitalist economic system. The task of integrating business has in some ways been aided by the trend to a smaller number of large firms. If Pilkingtons dominate the glass-industry, a fair number of other companies come close to monopolising their industries. Courtaulds and ICI not only control the artificial fibres industry but have large interests in the textile industry as well. The Lucas group is almost the sole supplier of a whole range of components for the motor industry. The motor industry itself, responsible for about 40% of British exports, consisted of about 13 sizeable firms in 1950: today it is down to four and one of those is struggling (Chrysler-Rootes). The heavy electrical industry has GEC, sugar has Tate and Lyle, cement the Portland Cement Company, textiles English Calico, and firms like Unilever have large fingers in a number of pies ranging from soap to ice cream (they have several common ingredients!). The list could go on until we had named 300 or so companies and by that time we would have accounted for a large slice of the national income.

This concentration is a perfectly logical development for competition is self-defeating. There is no league table in industry where firms get promoted and relegated but at the end of the season there are the same number of teams. The firms that get promoted eat the ones that get relegated.

A fact of life for every businessman is that the continued existence of his firm depends upon the elimination of as many as possible of those areas of uncertainty that face him. He is acutely concerned that the social, political, and economic environment is not subject to any marked change. A very natural concern, it must be said, in an age when industry is becoming more capital intensive. Equipment, unlike labour, cannot be laid off at little or no cost so it must therefore be kept running if it is to be profitable. But it can only be kept running if labour is there to run it, and if the market continues to exist for its goods. Labour must therefore be kept from striking, and the economic and political climate must be such that there are no violent fluctuations in the market.

As regards labour several policies might be adopted. Trouble may be bought off by putting almost any price (within limits naturally) on the compliance of labour, by establishing 'good relationships' with the unions and employees, or by pushing for legislation. Attempts can be made to secure the stability of markets by co-operating with governments (hence businessmen's willingness to sit on such bodies as the NEDC), and by international expansion which is a way of protecting overseas markets and of spreading political risks. But in the short run the best way for a firm to protect its position is to buy up its competitors, to buy up firms that provide it with raw materials or components, and to try and control in some way or other the outlets for its product. As regards this last point two examples may be given – the motor industry exercises a fair degree of control over car retailers, and both Courtaulds and ICI have moved into the textile industry.

What this very brief and condensed sketch shows is that since the war there has been a growing concentration of economic and political power. Yet it is a concentration not without contradictions – as firms get larger and exercise some form of control over larger areas of the economy they are in a stronger position to resist government regulation. The trade union movement must also be regarded as an unstable element in this structure of power for not all trade unionists subscribe to the 'national interest' theory, which has as one of its components the view that employers and employees have common interests. Indeed, Britain's two largest unions, the Transport and General Workers and the Amalgamated Engineers, show signs of moving away from this position and reverting to the more traditional stance which has it that the interests of capital and labour, far from being harmonious, are in fundamental conflict.

While the trade union hierarchies have been drawn slowly but certainly into the national administrative machine, while they have become converted to the 'national interest' theory of economic relationships, they have at the same time moved further and further away from the realities of the shop-floor. For at the shop-floor level it is not always so evident that the 'national interest' bears any relation to the workers' interest. This is not to say of course that shop-floor workers do not subscribe to the 'national interest' theory as well. (Very often they do – the RFSC at Pilkingtons, in their telegram to Barbara Castle urging an inquiry, spoke of the damage to the economy that could result from their strike.) Their attitudes often are quite contradictory – other people's strikes are wrong, but their own are all right!

In view of the absorption process to which trade union leaders have been subject it is not perhaps altogether surprising that since the war the number of 'official' strikes has steadily declined and the number of 'unofficial' strikes has steadily increased. Neither is it surprising that so much of the criticism levelled at the unions has been couched in terms of their inability to 'control their members'.

The problem for the union chiefs has of course been that while *they* have been taken into the national power structure, their power in the unions has been eroded. As all commentators like to point out these days, there has been a shift of power to the shop-floor. Union leaders are of course not unaware of this development even if they do adopt a different stance towards it. The Transport and General Workers Union has over the last few years developed a practice of referring agreements back to the shop-floor for their approval before a final ratification with employers. The GMWU policy, as of 1966, was to improve communications with shop stewards by bringing them into frequent contact with national and local officials. Thus Lord Cooper felt able to boast to the Donovan Commission (*Minutes of Evidence No 48*) that he had achieved some success with his shop stewards at Fords in Liverpool by bringing them all together in London. Two and a half years later in February 1969 the GMWU lost its entire membership at Fords – the members, led by their shop stewards, had gone to the T&GWU and the AEF.

It is not easy to predict how the movement of power to the shop-floor is likely to develop for there are a number of movements afoot. There have, over the last few years, been a number of attempts by shop stewards in different areas to form 'Rank and File Committees', the main purpose of which seems to be the co-ordination of assistance in the event of strikes. There have also developed industry based shop stewards committees – on Merseyside shop stewards from Fords, Vauxhalls, and British Leyland have regular meetings. 'Combine Committees' are another favoured form of organisation – several attempts have been made by Fords shop stewards to organise regular meetings of their colleagues from the plants in different parts of the country. None of these organisations have anything whatever to do with the official trade union structures, although different unions adopt varying attitudes towards them. Some, like the Transport Workers and the AEF tend to turn a blind eye, while others, like the Electricians and the GMWU are openly hostile.

There are signs too that there is some rethinking going on amongst shop-floor activists with regard to tactics. The signs are perhaps no more than straws in the wind – they are as likely to blow away as land and take root. In September 1969 there was a proposed factory occupation in the GEC plants in Liverpool: it didn't come off. Earlier in the same year during the Ford strike a number of activists were saying: 'If we don't get what we want, we'll go in and take this place over.' That tactics such as these should have been seriously discussed and planned – as they were in the case of GEC – may of course be no more than an indication that student tactics of sit-ins have *temporarily* influenced rank and file thinking and that the idea of occupation is no more than a passing vogue. Against this view however is the undeniable fact that the old idea of 'workers control' has woken from its long sleep and is now widely discussed amongst militants. Again, this may be no more than a fashion even if there is a much wider debate going on about 'participation', a debate given the accolade of respectability by Mrs Barbara Castle who was Minister of Employment and Productivity, when she said: 'I think it's part of the new social ferment, the new change of thinking in society when people from top to bottom feel that they want a more direct say in government, starting with the control of their own environment where they work, study, teach, or administer. This is part of people's definition of democracy. The exciting challenge to any government is how to harness what could be a very healthy reaction to an orderly conclusion' (quoted in *The Times* 20th May, 1970).

TRADE UNIONS AND THE POWER STRUCTURE

The 'wider society' apart, on the industrial scene at least there are two contradictory trends: a growing concentration of power on the one hand as trade union leaders become absorbed into the apparatus of government, and as the control of industry passes into fewer hands. And a paradoxical growth of power on the other hand at the factory-floor level where the shop stewards operate, more often than not with popular support, with very few checks from the unions of which they are members.

This national scene was to some extent reflected in the Pilkington dispute for not only did a full-scale civil war develop in the union, but the cry for democracy was taken up by the Rank and File Strike Committee.

Chapter Two

A STRIKE DIARY, AND THE ROLE OF THE PRESS

Strike diary

The strike took almost exactly forty-eight hours to get fully underway in the six main plants in St Helens – from the afternoon of Friday, 3rd April, to the afternoon of Sunday, 5th April. It started in the Flat Drawn department of the Sheet Works over a wage miscalculation, and had spread partially to the Cowley Hill works by the early hours of Saturday, 4th April. On Sunday afternoon it had spread in a matter of an hour or so to the four remaining plants. By this time the strikers were demanding a £10 increase on basic rates of pay, a £5 interim settlement being the price of a return to work. A shop stewards' meeting on the Saturday morning decided to recommend a return to work. This was put to a meeting of the Cowley Hill workers on Sunday morning and rejected. It was put to another mass meeting for the Sheet Works men on Sunday afternoon – and rejected. Immediately afterwards several thousand men marched around St Helens to the other works and persuaded the other workers to join them. Pilkingtons reaction was swift: on the Saturday morning a spokesman was quoted as saying (*Lancs. Post & Chronicle*) 'The company has made its position clear to the officials of the union concerned, that it is prepared to meet the union through the official Joint Industrial Council as soon as the men return to work . . .'

The next sixteen days between Monday, 6th April, and Tuesday, 21st April, were full of action, much of which laid the foundations for what was to follow in the remaining thirty-three days. Pilkingtons continued to reiterate their refusal to negotiate until the men returned to work although they were in fact negotiating with the GMWU on Friday, 17th April, Sunday, 19th April, and Monday, 20th April, and the GMWU urged its men to return to work so that it could get down to negotiations.

The strike spread to seven plants outside St Helens – Doncaster, 6th April; Birkenhead and Glasgow Fibreglass, 7th April; two plants at St Asaph in North Wales, 8th April; Pontypool in South Wales, 9th April; and Glasgow Triplex, 14th April. Four other plants never joined the strike – one in Kent, two in Birmingham, and one in London. The motor industry was badly hit by glass shortages – only one firm, Ford, found a reasonable alternative supplier from the Continent.

Local GMWU officials tried, unsuccessfully, to persuade their national executive to make the strike official and told the St Helens men that their strike was 'official at local level'. There were four mass meetings at each of which the men overwhelmingly voted to stay out: at two of them union officials were shouted down when they urged a return to work. Two of the mass meetings were organised by the newly emerged RFSC. At one of the meetings Lady [Mavis] Pilkington put in an appearance and did the popular press a good turn – 'The wife of the millionaire boss of a giant firm was jostled and catcalled by angry workers when she paid a secret visit to a massed meeting of strikers today' (*Lancs. Post & Chronicle*, Wednesday, 8th April).

The GMWU conducted a postal ballot through the Electoral Law Reform Society – the result of which was never subsequently published. The ballot itself was the subject of some controversy – 'Rebel shop stewards told a mass meeting this morning that their union's postal ballot was rigged ... glassworker John Potter told the crowd today: "Hundreds of these letters have gone out to men who are not even in our own union. And many of our members have not been given the chance to vote. The whole thing is rigged. It is treachery. We have evidence that ballot papers have been sent to AEF men and members of the Plumbers Union ..."' (*Lancs. Post & Chronicle* Monday, 20th April).

Unlikely explanations started to make an appearance after

a dramatic front-page headline in the *St Helens Reporter* – SCOUSE POWER PUTS PUNCH IN BIG STRIKE (Friday, 17th April), and a remark by Lady Pilkington: 'I think it was started by troublemakers from Liverpool, who don't care about the company or its traditions' (quoted in *Sunday Telegraph*, 19th April). Allegations about other 'outside influences' were made after a meeting at which Mr David Basnett, a national officer of the GMWU, was shouted down: allegations probably not unconnected with the appearance of people from Liverpool selling the rank and file Merseyside fortnightly *Big Flame* which devoted most of its front page to the strike. (According to one report it sold more than 600 in less than an hour.)

The RFSC gave its first example of its unconscious flair for making news by organising a march on Pilkingtons' head office where negotiations were going on between the firm and the union, and demanding an audience with Lord Pilkington. The march was preceded by the banner of a long since defunct glassworkers union discovered in the attic of the GMWU offices. Three days later it was announced that Pilkingtons had offered to add £3 a week to gross pay.

The period of nineteen days between Tuesday, 21st April, and Saturday, 9th May, began on a note of some confusion and not a little acrimony. It ended with some clarity for by 9th May there could be little doubt that the union had totally lost control of the situation, and that the RFSC was firmly in control. In this period all the factories outside St Helens returned to work – the two plants at St Asaph went back on 22nd April as did the Birkenhead plant, Pontypool went back on the 23rd, Glasgow Fibreglass on 27th April, and Glasgow Triplex on 4th May.

There were four mass meetings, two of them official union meetings, two of them organised by the RFSC. At the first of the official meetings – which had been organised by the RFSC but adopted by the union – the £3 offer was made and rejected (21st April); at the second (Friday, 24th April) the union officials had another rough ride and were never again seen publicly at a mass meeting. After an RFSC mass meeting on Wednesday, 29th April, some 300 strikers marched on the union offices, invaded them, scattered papers from an upstairs room and engaged in heated argument with the branch secretary. Shortly afterwards the branch secretary talked of resigning: he didn't. At the fourth meeting, again organised by the RFSC, two women dragged a shop steward from the platform after he had urged a return to work: 'One fall, one submission . . . and a technical knock-out, put paid to Matt McGrath's solo bid to end a £2m strike. In an unscheduled bout it took two women cleaners just a few seconds to dispose of 16-stone Matt and his "go-back-to-work-speech"' (*Daily Mirror*, 4th May).

A sign of the times was a May Day procession organised by the RFSC, led by two men dressed as Lord and Lady Pilkington (in clothes bought from a charity shop in which Lady Pilkington served as an assistant), and followed by 'pall bearers' carrying a coffin inscribed: NUGMW – RIP.

Public allegations were made that there were 'redsunderthebed' and that the strike had been infiltrated by 'subversive elements': '. . . allegations of a hidden element that has kept chaos and violence boiling at Pilkingtons came from Mr Bill X (the firm's chief shop steward). He said: "This town is being held to ransom by subversive elements and I have it on good authority that people have been brought in specially to erupt trouble outside the various factories"' (*Daily Express*, 9th May). This was denied by 'Brian X . . . a member of the unofficial strike committee . . . "Maoists?" he says. "If you can find half a dozen copies of 'The Thoughts of Chairman "What's-his-name"' in St Helens I'll show my backside in Woolworths' window on Saturday morning. This is a good Catholic town, and good Catholic lads don't go for Communism"' (*Daily Express*, 13th May).

The RFSC attempted to negotiate a settlement through the Mayor and two local MP's (Sunday, 26th April). The MP's put forward a peace formula of which nothing further was heard publicly. On Monday, 4th May, the RFSC cabled Mrs Barbara Castle, Minister of Employment and

Productivity: REQUEST INTERVENTION BY YOUR DEPARTMENT INTO THE PILKINGTON GLASSWORKERS DISPUTE STOP ONLY BY YOUR INTERVENTION CAN THERE BE AN EARLY RESUMPTION OF WORK (quoted in *St Helens Reporter*, 5th May). On Saturday, 9th May, Mrs Castle announced a Court of Inquiry into the dispute – partly because, according to John Torode (*New Statesman*, 15th May), the Department of Employment and Productivity had been persuaded that allegations of 'outside intervention' were of some substance. (The Court of Inquiry, in the event, never so much as attempted to examine the 'red scare'.)

Both Pilkingtons and the GMWU attempted, unsuccessfully, to engineer a return to work – with the exception of 150 men working at the Pilkington research laboratories just outside St Helens. On Monday, 27th April, Pilkingtons sent a letter to all their employees explaining the terms of the £3 offer and urging a return to work; on 30th April they extended the £3 pay rise to all adult employees not on strike and claimed that they were being 'inundated with requests from workers wishing to return to work'. The firm said that they were ready to take them back and give them full protection. On Monday, 4th May, the RFSC said that they had proof that members of Pilkingtons' welfare staff had been calling at the homes of widows and wives of staffmen saying that if they did not return to work they would be sacked. Pilkingtons declined to comment.

The GMWU managed to organise small factory gate meetings (*inside* the gates with Pilkingtons' permission) on Friday, 1st May, and Saturday, 2nd May – at five out of the seven they secured a vote to return to work. At a mass meeting on the next day the RFSC mustered 4,000 people who voted to stay out.

There was a partial return to work on Friday, 1st May, Monday, 4th May, and Tuesday, 5th May. By Wednesday, 6th May, it had petered out after mass pickets at one of the works gates had successfully kept returning workers inside the gates at knock-off-time until they agreed not to return to work the next day. The popular press had a field day over this three-day period by announcing violence in banner headlines, illustrated with appropriate photographs: 'SIEGE AT TEA TIME!' yelled the *Daily Express* on Wednesday, 6th May. 'EIGHT HUNDRED screaming, punch-throwing pickets ambushed 385 workers leaving a strike-hit glass plant at teatime yesterday. Within minutes violence had exploded into a riot and a siege. The militants flung themselves at police guarding the factory after the employees had retreated inside for safety... No workers turned up for night shift.' There was in fact very little violence although there were some very angry scenes in which there were a few minor scuffles, and abuse and pennies were hurled at the strike-breakers.

The most important event of this period of the strike was missed by the press: the confused meeting of all shop stewards at the GMWU offices on the morning of 21st April, a few hours before the mass meeting at which the £3 offer was recommended as a suitable basis for a return to work. Union officials and some shop stewards left that meeting with the impression that the shop stewards were unanimously behind a *recommendation* for a return to work. But members of the RFSC left it with the impression that it had been unanimously agreed that the offer would be put to the mass meeting for *their* approval without any recommendation *either way*. It was because of this misunderstanding that union officials were subsequently to make allegations of treachery about certain members of the RFSC.

At the beginning of the final stage of the strike – lasting from 10th May to 22nd May – it looked as if the strikers were digging in for the winter. Attempts to get a return to work had failed, the firm gave positively no indication of being prepared to budge, the GMWU was blundering and floundering around sponsoring wild and scatty statements by some of its shop stewards, and the RFSC seemed to be unquestionably in charge of the situation.

Probably more significant than the return to work on 22nd May, the Friday of Whit weekend, were events at the

A STRIKE DIARY, AND THE ROLE OF THE PRESS

mass meeting at the beginning of the period. On Sunday, 10th May, the RFSC distributed forms to the crowd: the forms instructed Pilkingtons to stop deducting union dues from their wages when the signatories returned to work. The best part of 3000 were signed and handed back to the RFSC. At the same meeting a resolution was carried unanimously to the effect that there would be no return to work until twenty-eight men at the Pontypool factory had been reinstated. There were six further mass meetings.

Union representatives continued their allegations of 'red subversion', although two days after an hysterical outburst about Maoist influence at a specially convened press conference, a GMWU official said: 'Frankly, there is not a shred of evidence' (*Daily Sketch*, 14th May). In the sixth week the GMWU paid out a hardship allowance of £12 to men and £6 to women – which was promptly called a bribe by the RFSC because the strike continued to be unofficial. In the middle of the same week – on Wednesday, 13th May – the RFSC agreed to a confrontation on Granada television with GMWU representatives. During the course of the programme the GMWU announced that it was organising a ballot for a return to work on Saturday, 16th May, to be run by local clergymen of all denominations. The RFSC countered by asking if the GMWU would be prepared to also run a ballot to see how many people wanted to stay in the GMWU. The RFSC did not push the point and eventually agreed to participate in the organisation of the 'parsons' poll' on the issue of a return to work.

The ballot showed a 4% majority favouring a return. The next day, Sunday, 17th May, the RFSC called a mass meeting, alleged irregularities in the conduct of the ballot, and between 3,000 and 3,500 strikers voted to continue the strike. The RFSC also thought it had lost in the poll because it knew that some of its supporters, confident in the result, had gone to the Rugby League final in Bradford without voting. The next morning, Monday, 18th May, Pilkingtons claimed that some 2,500 workers had reported for work – a figure hotly disputed by the RFSC who said that staffmen had gone into work with the strike-breakers and thereby inflated the figures. The police put on a show of force at the factory gates for the first three days of the final week. Although there was no 'trouble' when people went into work in the mornings, there were confrontations between pickets and workers in the evening. Three arrests were made on the Monday, and eighteen on the Tuesday. The first three were given prison sentences – one of the sentencing magistrates subsequently had his shop windows smashed.

The Court of Inquiry opened in Liverpool on Tuesday, 19th May, but did not seem to arouse much enthusiasm in St Helens or in the RFSC. The RFSC did eventually give oral evidence but ignored appeals for a return to work. Pilkingtons continued to be fairly silent in public. On Wednesday, 13th May, strikers received a letter from the firm warning of lay-offs if the strike continued and reminding them that it was a condition of employment that they should be members of the GMWU (letter reproduced in the Appendix). Pilkingtons only other important action was to arrange for a large number of police to be available when some of the strikers returned to work in the last week.

The strike finally ended with the RFSC's acceptance of Vic Feather's (TUC General Secretary) offer of mediation between them and the GMWU.

What some of the papers said

The press generally was not popular with any of the 'generals' during the strike. But where GMWU officials did not praise *any* of the papers Pilkingtons and the RFSC were more discriminating – both liked *The Financial Times* and both disliked the popular press. In this section, as a prelude to a more general discussion of the press, we shall compare *The Financial Times* with the *Daily Mirror* with regard to the way they reported several important events.

On Tuesday, 21st April, a mass meeting of strikers rejected an offer of £3 per week to be added to gross pay. The following morning the *Daily Mirror* headlined a 230-word piece: GLASS WORKERS REJECT £3-A-WEEK OFFER. *The Financial Times* headlined a 400-word piece: PILKINGTON STRIKERS REJECT OFFER OF £3 RISE. Both stories gave a fair bit of space to the effect of the strike on the motor industry – about one-third of each was devoted to the tribulations of Rootes and Jaguars. The coverage of the mass meeting however was quite different: where the *Daily Mirror* reported the meeting as being a straightforward rejection of the £3 and ending with an overwhelming decision to continue the strike, *The Financial Times* reported some confusion. The *Daily Mirror* reported '. . . union negotiator David Basnett . . . (as being) snubbed', *The Financial Times* said: '. . . militants shouted down Mr David Basnett . . . and bombarded him with paper darts when he proposed that work be resumed.' *The Financial Times* also noted 'But several hundred workers supported this recommendation only to be jeered and outvoted. It was reported that many did not vote at all and after the meeting there were claims that the vote was only marginal and that many had not heard the vote called.' *The Financial Times* reported too (which the *Daily Mirror* had not) that: 'Earlier in the day shop stewards had voted unanimously to accept the peace formula drawn up by their union negotiators . . .'

The overall tone of both reports was sober and unemotional, but *The Financial Times* story revealed some of the complexities of the situation while the *Daily Mirror* reported it as quite clear-cut. It was left to *The Times* to report (along with the *Daily Mail* and the *Guardian*) that the strikers at the St Asaph plant had accepted the terms of a return to work, and to the *Daily Telegraph* and the *Guardian* to report the differences between the positions of the RFSC and the GMWU at the meeting. No one newspaper carried a full story, but only one – the *Daily Express* – carried a 'sensational' headline: NIGHTMARE STRIKE.

In the strike that provided the press with more than one bonanza, the first big bonus was paid out on Wednesday, 5th May, when the GMWU and Pilkingtons made their first attempt to get a return to work. The following morning

WHAT SOME OF THE PAPERS SAID

the *Daily Mirror* billed its fare: STRIKERS IN BIG CLASH AS SHIFT ENDS. *The Financial Times* noted POLICE CALLED AS PICKETS THREATEN PILKINGTON SHIFT with its story about one-third longer than the *Daily Mirror*'s. Half of the *Mirror*'s revolved around 'Police reinforcements were rushed in as violence flared.' *The Financial Times* by comparison devoted about a quarter of its story to the same theme. Although both papers presented much the same story, the *way* it was presented was quite different – the *Mirror* led off its piece: 'UNOFFICIAL strikers fought with workers outside a factory last night.' *The Financial Times* led in with: 'FIFTY ... police were called ... yesterday as men who had returned to work were threatened for more than an hour by pickets at the factory gate.' *Daily Mirror* violence was 'fights'; *The Financial Times* violence was 'taunts' and thrown pennies. The remainder of both stories was also quite different: *The Financial Times* reported that the RFSC had successfully blacked gas supplies and had cabled Mrs Barbara Castle seeking her intervention. The *Mirror* reported neither of these points but did say that the GMWU had ordered an inquiry into the local branch – which *The Financial Times* did not report. Once again *The Financial Times* gave a fuller report, but it could not have been said on this occasion that the *Mirror* was sober. Its story was not, strictly speaking, inaccurate for there were as a matter of fact several scuffles but the impression was nevertheless created that there was fighting on a large scale. Sentences such as: 'Police reinforcements were rushed in as violence flared' suggested a sense of urgency and confrontations of some magnitude – a sense that the situation could not have justified. But the *Mirror* was mild compared with the *Daily Express*.

While the *Mirror* buried the story on page eleven, the *Express* splashed it across the front page –

Battle of Grove Street
SIEGE AT TEATIME!

and led in with: 'EIGHT HUNDRED screaming, punch-throwing pickets ambushed 385 workers leaving a strike-hit glass plant at teatime yesterday. Within minutes violence had exploded into a riot and a siege.' A frankly quite incredible first paragraph considering the nature of the events. The 330-word piece, 80% of which was given over to 'violence', was accompanied by two large photographs one of which took up twice as much acreage as the actual story, the other occupying half as much space as the story. *The Times* report carried no mention of violence but *The Guardian* gave it some prominence. *The Daily Telegraph* devoted a small piece to it on the front page but a much longer story on an inside page devoted half to a statement from the RFSC. The fullest coverage was again given by the 'heavies'. Yet again no single paper carried the full range of relevant information.

Violence was once again the big news at the beginning of the last week of the strike. *The Financial Times* on Tuesday, 19th May said: 'VIOLENT SCENES MAR PILKINGTON RETURN', and followed up with a 500-word piece. The *Daily Mirror* covered its centre pages with photographs captioned: 'Trouble flares outside one of the Pilkington factories . . . Policemen scuffle . . . with an angry picket who has lost his head in the heat of the moment. But the man on the motor-bike coming out of the factory gates doesn't seem to want to know about it. He appears to want to get quietly away after his shift in the glass plant. As police and pickets battle out their shirt-sleeved duel in the sun, he slips discreetly past . . .' The *Mirror*'s wordage that day was headlined: ELSIE GETS JEERS FOR DOING THE RIGHT THING, a 370-word story being devoted to how a woman, a former member of the RFSC, had decided to go back to work. Another *Mirror* story that day (380 words) was headlined: VIOLENCE FLARES AS GLASS STRIKERS GO BACK. Excluding advertising space, the *Daily Mirror* devoted the best part of three pages to the strike, but only half of one of those pages reported any detail of the day's events. As far as coverage was concerned the *Mirror*'s reporting was almost as full as *The Financial Times*. The only point missed by

WHAT SOME OF THE PAPERS SAID

the *Mirror* but not by *The Financial Times* was the imminent TUC intervention.

All of the other dailies devoted large expanses to print and photographs, but if *The Financial Times* used least space it reported all there was to report.

The rather brief examination of the reportage of *The Financial Times* and the *Daily Mirror* of three different events shows two things very clearly. First, that when it comes down to pay-dirt, to facts, there was very little to choose. Second, that the *presentation* of news is what marks one paper off from another. Presentation includes such things as the choice of headlines and sub-headings, the use of bold type and photographs, the style of writing and the order in which events are written up. There is an obvious difference in the style of presentation: the 'pops' (*Daily Mirror, Daily Express, Daily Mail, Daily Sketch, The Sun*) go for the emotional and arresting headline, the 'heavies' (*The Times, The Guardian, The Daily Telegraph, The Financial Times*) for the sober and more objectively descriptive. The 'pops' style of writing is jazzier and goes for the colourful and evocative adjective.

None of the papers were *openly* hostile towards the strikers, and needless to say none of them were favourably disposed either. The firm came in for some criticism – it was called 'paternalistic', and so did the GMWU – it was said to be out of touch with its members. The RFSC however was more difficult to pin down for it could not have been held responsible for the start of the strike, and it was difficult to blame it for the continuation of the stoppage because there were so many mass meetings at which several thousands of people regularly voted to stay out. One paper, *The Guardian*, came very near to supporting the strikers – without actually expressing support in so many words. Several, the *Daily Express*, the *Daily Sketch*, and *The Sun*, came very near to open hostility without actually expressing it either.

Yet despite the leanings of some reporters, despite the editorial policies of the different papers, all of the press without exception managed to convey the impression that the whole thing was really rather lamentable. Certainly some lamentations were more subtle than others, certainly praise and blame was apportioned more carefully by some papers than others, certainly some papers tried harder than others to be neutral. But the odd phrase or word showed through the most patiently worded piece to render the most neutral less than neutral. The industrial status quo, in other words, always came out on top. This of course should not surprise anyone for the press itself is part of the status quo, but it does suggest that the idea of a 'free press' has rather less substance than its proponents would have us believe. No doubt many newspaper readers would be rather shocked to find their paper saying what a *good* thing strikes were, but such a point of view, though quite as legitimate as the reverse, never makes an appearance. Strikes are never written about from the point of view of the striker. It may be that from time to time a pay claim is seen as justified and that the press is unanimous in this view, but a strike in its pursuit is nevertheless regarded as a matter for regret. A more balanced press would contain a few widely read papers rejoicing in strikes as subtly or as crudely as the existing press deplores them. As the press is presently constituted there is not much joy in it for the striker – the most he can hope for is the strained neutrality of papers such as *The Financial Times*.

The press and the 'public'

A strike at Pilkingtons in St Helens is not the same as a dock strike in either Liverpool or London, for a strike in a smallish town whose labour market is dominated by one firm affects everyone. If only about one quarter of the working population works at Pilkingtons, almost everyone in St Helens either has a relative, a friend, or an acquaintance working there. In addition, at least 80% of those who work in the firm live in, or very close to, St Helens. St Helens is compact: virtually everybody lives within two miles of the town centre and one or other of the Pilkington plants. St Helens is also richly populated with pubs and Labour clubs. The town, in short, had all the necessary ingredients for an effective informal communications network. Testimony to this was the speed with which the RFSC could organise mass meetings of considerable size at a few hours' notice. Unload into this situation the concentrated resources of the national and local press, national and local radio and television, and you have a town living on a staple diet of strike.

In a strike, where the war is waged with words but between people who are not speaking to each other, the press assumes a crucial role. It becomes crucial because it carries, so to speak, messages from one party to the other and simultaneously gives a picture of the current state of play to the public at large. But the press was more than a 'despatch rider' lathering from Ghent to Aix (or from Pilks head office to the RFSC HQ in the Cotham Arms), for the despatch rider had a mind of his own. It would not, therefore, be going too far to say that since the press was the main bearer of first hand news, the way in which it presented its news had very real consequences for the way people thought and hence for the course of the strike.

One particular aspect of press coverage particularly incensed both Pilkingtons and the RFSC – the massive coverage given to 'violence' on the picket lines in the fifth and seventh weeks of the strike. Both were incensed because they thought it to have been blown up out of all proportion – though for different reasons. The firm because they thought it deterred a lot of people from returning to work; the RFSC because they thought it brought discredit upon them and branded them as 'thugs' and 'hooligans'.

There might well have been something in Pilkingtons' view. In the sixth week when we conducted our survey we asked all those who favoured a return to work (54%) why they had not done so. Well over half of these (63%) gave reasons such as 'fear', 'afraid of violence', 'afraid for my family', etc. Some eight days previous to these answers the

Daily Mirror (more widely read in St Helens than any other national daily) was producing headlines: STRIKERS IN BIG CLASH AS SHIFT ENDS (6th May) and STRIKE-BREAKERS BEATEN BY FEAR (7th May). These headlines accompanied by introductory paragraphs in similar vein, and with the subsequent story told in vivid language presented an image of no little chaos and of personal risk to those who had returned to work.

Now we do not know that these accounts were responsible for subsequent expressions of fear on the part of our respondents, though we may wonder what responses these same people would have given had there been no excessive publicity given to 'violence'. Up until that week there had been no 'violence', and as a matter of fact 'violent' was rather a strong word to describe occurrences in the fifth week. There were no arrests, and if there were a few punches thrown and kicks delivered, they were not thrown or delivered indiscriminately.

Indeed the trouble was largely centred around one man who had a lucrative part-time job. This man had gone into work tapping his hip pocket, jeering at the pickets, and signifying 'fuck-off' with two fingers. There were inevitably several of the pickets who singled him out for that evening when he knocked off. The pickets did of course jeer at the strike-breakers and they also threw pennies, but the 'violence' was basically of a *moral* kind. Strikes are not won if there are breaks in the ranks and this is always well understood by strikers. The whole purpose of picketing is to try and shame those who want to go back to work to stick by their fellows. Not for nothing has 'united we stand, divided we fall' been the great rallying cry of labour movements the world over. In the event 'extreme moral pressure' rather than 'violence' would have been a more fitting description. This amounts to saying that the views of our respondents were out of all proportion to what had actually taken place, and it seems reasonable to suspect that some responsibility for these views must rest on the popular press and television. Apart from this one instance of strong inference we cannot hang anything on the press which refers to specific instances.

In the last resort it is quite impossible to decide just how important the mass media are – if only because people are not empty receptacles into which information can be poured. News stories are interpreted according to how much information the reader already has, according to his contacts with other people, according to whether or not he has first-hand experience and knowledge of what is being written about. Nobody in St Helens read a news story 'cold' – the strike was too close to home for people not to have developed some sort of mental translating equipment. In those circumstances there was no chance of any story being read uncritically, and every chance that the interpretation inherent in every story would be rejected by someone or other.

Furthermore the strikers were not – as will be pointed out in the next chapter – a homogeneous mass. They were not all of one mind at any one moment in time, so that there could have been as many interpretations of a *Daily Mirror* piece as there were positions amongst the strikers. Thus for those on picket at the time of the 'violence' the *Daily Mirror* will probably have appeared as a gross piece of sensationalism, while for those who wanted to go back to work the same piece may have served to reinforce doubts already held about running the gauntlet of their mates' disapproval and making the act of strike-breaking appear doubly dangerous.

We have said that the strikers were not an undifferentiated mass; to be fair we should also point out that the press was not an undifferentiated mass either. If it can be said at one level that the 'press' is one of the important buttresses of the established order of society, such a statement glosses over many of the subtle distinctions in the world of the press. The industry – and we need to remind ourselves more often that it *is* an industry – is competitive, and not just in the matter of circulation battles, for reporters vie with one another to get that little bit extra that makes their story that much better. Different papers have their own eccentricities and political leanings, and editorial policy is not necessarily a reliable guide to the ways in which stories are written. Right-wing papers have left-wing reporters and vice versa. Some papers edit reporters' copy more than others (or at least that is what reporters tend to tell people who are not pleased with a story's presentation); some reporters are more prone to sub-editing than others; some papers sub-edit more than others; some reporters are more sympathetic to one side or another than other reporters. Indeed during the strike different reporters writing for the same paper leaned one way or another to such an extent that the apparent 'line' of the paper could vary from day to day.

It might finally be pointed out – for the benefit of those predisposed to hostility towards reporters – that there is a wealth of meaning in the reporter's disclaimer: 'I've got a job to do just like everyone else.' The reporter is saying, amongst other things, that he is not a free agent, that he cannot write exactly what he wants, that he is as much constrained by the system of power and authority in his newspaper as is a glassworker at Pilkingtons.

The press and the 'generals'

The 'generals' – the RFSC, the GMWU, and Pilkingtons – were probably much more selective in the way they looked at the press than was the rank and file striker – they read the 'pops' to get some indication of how the 'general public' were being informed, and read the 'heavies' for their own information. The 'heavies' were relied upon not only because they tended to give fairly full accounts of what the parties had said and done, but also because they usually presented the story in an unemotional manner. The 'heavies', therefore, when taken into consideration with the supposed influence of the 'pops' on the general public, provided some basis for thought and further action.

But if the generals *reacted* to the press, they also played an active part in determining what went into it. They did this in broadly two ways. They could call a press conference and make statements which might determine the terms of debate that day by forcing the other side to issue a denial or to counter with another move, or they could use the press to publicise a move designed to secure public sympathy.

The 'heavies' almost certainly exerted more influence on the RFSC than on the other two parties – but through its feature articles rather than its reportage. Several RFSC *Strike Bulletins* quoted extensively from *The Guardian* features: in this case using quotations as an authoritative source of information to back up its own claims.

But feature articles were most influential in another and more indirect way. Feature writers are given a licence not ordinarily available to journalists: a licence openly to analyse and interpret material and to come to a conclusion. (Of course all journalists interpret material but they have to do it more surreptitiously.) Which is another way of saying they are permitted to apportion praise and blame.

A reading of the 'heavy' features by the RFSC allowed them, with some justification, to conclude that on the whole they came out of it rather well. Or if there was criticism, and there was, they tended to get off rather lightly. Pilkingtons did rather less well: a *Guardian* feature said on one occasion (4th May) 'Pilkington would certainly refute any suggestion of strike-breaking but this is what the company did as effectively as if it had brought in substitute workers.' The GMWU came off worst of all for hardly any commentator at any time could find a good word for it. Little wonder that Lord Cooper was moved to comment: 'I am sick and tired of Press and TV reports based on information which is wrong and false ... The organs of publicity do not contribute to better relations in industry ... (The press has) damaged the name of the union and made things worse for us ...' (quoted in *The Financial Times*, 18th May).

The feature writers in the 'heavies' were usually experienced industrial reporters not given to 'flights of fancy': their interpretations were always based on solid evidence and never stretched it beyond the limits of credibility. If we did not always accept their judgements we had at least to accept that they were reasoned. The RFSC took a similar view and consequently took them seriously. They were prepared to admit – in private – that they were not without fault, but the fact that the other two groups of 'chiefs of staff' seemed to be getting more stick than they, tended to reinforce their feelings that right was on their side. The measured, objective tones of features only added to this conviction: here were experienced, unsensational, and apparently impartial reporters 'giving us a fair crack of the whip'.

There was one other way in which the features were influential: their analyses went beyond the immediate situation of the strike and attempted to put it some sort of historical perspective. They latched on to the traditionalist ideology of Pilkingtons and to the idea of St Helens being a company town. They noted that St Helens was just down the road from Liverpool, the scene of some notable militancy in recent years and that St Helens people could hardly be unaware of what had been going on, particularly in the motor industry. They pointed out that the GMWU was hardly noted for its militancy, that indeed it was on any criterion easily the most conservative of the big unions. The RFSC read all this and thought it made a good deal of sense. It helped them to clarify many of their own ideas, it provided them with some sort of framework into which they could fit many of their own experiences. Most important of all it helped them to think much more clearly about all that was involved in their enterprise.

Part Two

Chapter Three

THE RANK AND FILE

The unknown but decisive factor

In most strikes the attitudes of the rank and file participants are an unknown factor. Such was the case during the Pilkington dispute. No reliable information was available about the views of the rank and file strikers towards the stoppage. The only available indications of their feelings were of doubtful validity. Ballots and mass meetings provided some information but the often complex attitudes that individuals possess can rarely be adequately expressed in response to a simple motion at a mass meeting or the single choice given on a ballot paper. The participants in the strike themselves had no way of knowing exactly what the views of the rank and file were. Indeed, the state of rank and file opinion itself became the subject of a fierce debate in which various groups contending for leadership claimed to be representing mass opinion.

Although it is an unknown factor the rank and file must play a critical role in any dispute, because only its decisive action can start and terminate a strike. For a strike to begin and end the rank and file must act decisively. In the final analysis a strike consists of the behaviour of its rank and file participants.

Because it plays a critical role during a strike, a full understanding of the course of a dispute requires that the mystery of rank and file opinion should be resolved. In order to unravel its complexities, we conducted a survey in St Helens during the sixth week of the strike, consisting of interviews with a sample of 187 strikers. In designing this survey, we were handicapped by not having at our disposal any sampling frame from which a random sample of strikers could be drawn. What we did was to select one district in St Helens and by means of a household canvas we contacted and interviewed 187 strikers who were living in the area. The strikers who lived in the particular district that we chose for our investigation may have been untypical, although we have no concrete reasons for believing that they were. Each striker was questioned about his attitudes towards the strike and towards the various parties who were involved in it. The questions were simple and the answers were pre-coded. This type of interview survey can never hope to do more than scratch at the surface of individual's views and opinions. Nevertheless, it is the most effective method available of assessing the opinions of large groups of people, and in interpreting the results obtained from the survey we have been able to draw upon the insights we obtained in the process of carrying out other parts of the wider study of the strike. In this chapter we shall use the information obtained from our survey of rank and file opinion in order to present a picture of the manner in which the rank and file influenced and was influenced by other groups and events during the course of the dispute.

The outbreak of the strike

When we embarked upon our survey we expected to discover considerable bitterness on the part of the strikers towards the firm with which they were ostensibly in dispute. The press had been featuring the strike as an extremely bitter and at times violent struggle. Furthermore, previously formulated theories offering explanations of the outbreak of wildcat strikes had suggested that often they occur as explosions of accumulated grievances and ill-feeling.

However, the results of our survey revealed that on the whole the strikers' attitudes towards their firm were positive rather than critical. In response to an open-ended question on the things that were wrong at Pilkingtons before the strike, 59% said that levels of pay were generally too low. When asked a specific question about levels of pay in their firm 67% expressed the view that they were below the average throughout British industry as a whole. But apart from this issue of pay few strikers mentioned anything else as being particularly wrong with Pilkingtons. 19% criticised the union and negotiating machinery, 10% complained that their earnings tended to fluctuate from week to week, and 10% also criticised anomalies in the pay structure which resulted in considerable differences in the earnings of individuals doing very similar jobs. Quite a list of grievances was mentioned by all the strikers taken together. But few individuals criticised anything other than their overall levels of pay.

The general condition of rank and file opinion was one in which few individuals had many criticisms to make of the company. The rank and file displayed little hostility towards their firm despite having been on strike for over five weeks when the interviews took place. The atmosphere in the firm was not seen as being unduly bitter and hostile; 69% chose to describe the state of labour-management relationships in the firm in terms of 'everyone working together as a team', whereas only 23% thought that 'repeated conflicts' was a more appropriate description of shop-floor relationships. Towards their own particular jobs the strikers possessed mostly favourable feelings. 81% said that they would rather be at work than at home, even if it made no financial difference to them. Little bitterness was felt towards Lord Pilkington, the chairman of the company and generally recognised as being the man in charge. 92% strikers said that they felt no hostility towards him whatsoever.

Against the background of their self-confessed attitudes towards their firm, one might wonder why the strikers had become involved in the dispute at all. Indeed, this was a question that puzzled the strikers themselves, for 80% claimed to have been surprised by the outbreak of the strike, and 80% also claimed to have been drawn into the dispute reluctantly. The majority of the Pilkington employees were not actually at work when the strike broke out. Only those employees who were on the relevant shifts during the critical weekend were involved in a walk-out. Others simply heard that a strike had begun or were informed by pickets that a dispute was under way when they next reported for work. From their own accounts, most of the workers evidently did not respond with any enthusiasm to being told that they were on strike. Many of those who had actually 'downed tools' said that they were acting with reluctance. They explained their involvement in the strike as a response to advice from shop stewards or pressure from their mates. A minority (20% according to the results of our survey) welcomed the strike and were behind it from the beginning. But the behaviour of the majority of the strikers was not a direct response to their own feelings and attitudes; they became involved in the dispute in response to immediate and unanticipated situational pressures. Indeed, 45% revealed that when they first became involved they had no idea what the strike was about.

It may seem difficult to understand how thousands of workers could become involved in a strike that they had not expected, that they did not support, and whose objectives were obscure. The following accounts of the outbreak of the dispute, given to us by eye-witnesses, may help to make the behaviour of the strikers comprehensible. Firstly, we have the recollections of an employee from the Flat-Drawn department at the Sheet Works where the trouble initially began.

'This week they were paid short again – they hadn't been paid according to the agreement although management maintained they had. This was it. There was a terrific amount of ill-feeling.

'I went into the office with the shift shop steward to see the manager, and he assured us that he would do all he could to see it straight. We thought this was reasonable, but the resentment was there amongst the men because it hadn't been straightened out the week before. Anyway, we went back upstairs to tell the men and they said: "If it isn't straight

THE OUTBREAK OF THE STRIKE

by mid-day, we're going home." We then went back to the manager and told him what they had said. He said: "We've done all we can," which was fair enough – he couldn't do any more, his hands were tied.

'So we went back and told the blokes this, but by the time we got there – about 11.30 by now – the men were off the machines with a manager there with them. He was trying to talk them back, but they weren't having any. At this point, we tried to talk them back as well, but they still wouldn't have it. Then the manager said he'd meet a delegation of ten or twelve in the office, so we put this to them and they said: "Aye, all right."

'Meanwhile, before the deputation went in, they'd started talking about the half-a-crown an hour which had been a sore point for a long while because this particular shift had urged the union branch to go for 2/6d on the base rate, but had been turned down by the branch. So they decided to call this a strike for either 2/6d an hour, or a consolidation of bonus into the base rate. At this point they decided to go out, and by this time word had spread like wildfire, from the top floor to the bottom and across to No 6 tank four or five hundred yards away.

'The deputation to the management did take place but not straight away. They walked off the job and met in The Bridge (a pub outside the gates) to pick the delegation – five from Nos 6, 8, and 9 tanks.

'When we eventually got in the office Pilks had got hold of one of the union officials, and I'd made desperate attempts to get in touch with another official but he was at lunch. The senior shop steward had also been sent for, and when he came over, we put him in the picture and told him that it was now for 2/6d an hour. Then the union official came over and said in his usual manner: "We can't do this here, this has got to go to the JIC, you can't do this at local level. You're wasting your time putting it to management. Let's have a return to work and sort it out at JIC level." The men told him where to go.

'Then management wheeled in. They expected to be

tackled about bonus and came in with a big set-piece explanation, but when they finished one of the blokes said: "You're wasting your time talking about that – it's 2/6d an hour on the base rate, that's what we want." One of the managers turned round to the union official and said, with amazement written all over him: "What's all this?" The union bloke replied: "I don't know, I'm as wise as you are. I've just come in and they say it's for 2/6d an hour." Anyway, the manager said there was nothing to be done at local level, but as regards the bonus issue, they'd have men working all weekend to sort it out. And of course, there *wasn't* anything management could do about 2/6d an hour on the spot.

'The deputation then voted, about 17 to 3, to recommend staying out, and they put this to a meeting of about 350 blokes in the canteen. They took a vote and decided to stop out. Then they started shouting and chanting: "Let's get Rolled Plate (another department of Sheet Works) out," and one thing and another. This was about 3.15 pm because the deputation to management dragged on for about three hours. About 150 of them went over to Rolled Plate then and there, told them what it was about, and persuaded them to come out.

'Those of us left in the canteen had never been in a strike before and didn't know what to do. They were talking about pickets, but eventually it was left to an individual decision – Go and picket if you feel like it. They decided to send pickets up to the other plants to try and get their support, but we couldn't get any organisation into it at all.

'So it finished up with about six of us going up to Cowley Hill that night – about 9.30 pm – because we knew that if we were going to stop Pilks, we had to stop Cowley Hill (the float glass plant). We spoke to the shift going on and asked for their support. Some didn't want to know us, some said: "Yes, we'll stop out now and stop the others going." But we told them not to do this, but to go in and let the afternoon shift off, not to start work and to try and persuade the others not to start – which they did, and did very effec-

THE OUTBREAK OF THE STRIKE

tively. We spoke to the afternoon shift coming off at 10 pm, but they didn't want to know; they were after a bevy before the pubs closed at 11 pm. Quite a lot were in cars and of course we couldn't get to them.'

From here, another eye-witness from the Cowley Hill Plant takes up the story.

'It was a normal Friday evening as far as I was concerned. I went into work at about 9.20 pm and spent half an hour in the canteen with 20 other blokes: we talked about all the usual things but there was no talk of a strike at Sheet Works. I went into the department about ten and started work – not five minutes had passed when a group of about 40 men came up and said: "Why are you working? – we're out on strike!" I had a quick glance at the crowd but I didn't see a shop steward so I finished my job, which took about five minutes, and then talked to the men. They said they'd spoken to the Sheet Works pickets outside and that they were going out in sympathy. They didn't seem to have much idea of what was going on – of why the Sheet Works men were out, so I asked them to wait while I went to see if I could find a shop steward.

'I found two stewards not forty feet away, one of whom had talked to the pickets who had told him it was a bonus issue and for 2/6d an hour. I said: "What are you going to do?", and they said: "We're calling the men out." By this time, the crowd had grown to about 75 men and we all went outside the gates waiting for a meeting to be organised of all the night shift.

'After half an hour a bloke came out and said everyone was meeting in the canteen. We sent six men off to the meeting but the security men on the gate stopped them. Fortunately, four of them got through anyway, one of whom came out to tell us that about 100 men were meeting in the canteen. Then a shop steward and a personnel man came out and said we could all go into the canteen to talk about the problem.

'When we got into the canteen there were tables ranged at one end and behind them were sitting four department managers and two personnel men. The works manager said: "Let's discuss the problem." Then suddenly everyone seemed to cotton on at once that it wasn't a union meeting as they'd thought and started to chant: "Out, out." The managers went out rather quietly and sheepishly and stood around outside the doors, but we posted six big lads on the door.

'Then we sat down, drinking tea – about 150-200 of us – and waited while six shop stewards sat down in a corner to talk. With them was a man from Sheet Works who somehow or other had got in. Eventually one man got fed up waiting, went up to the stewards, and asked them what was happening. One of the stewards said: "Out, out of the way, this is nothing to do with you. We'll tell you what we want you to do later." This chap replied, in a loud voice: "They're selling us up the road again." But at that point, the stewards broke up and said that as far as they were concerned, there was no dispute and that we should start work. There was a lot of shouting then, so the stewards stood the Sheet Works bloke up on a chair and asked him to tell us what it was all about. This was a disaster because he just didn't know the first thing of what it was all about. He really didn't know a thing.

'Then one of our blokes got up and said that he'd been talking to another Sheet Works bloke outside, and he went on to explain what it was about. Then another bloke got up and said we'd got to decide whether to stop in or go out. Another bloke stands up, calls for a vote, and all the hands go up to go out. Then the stewards tried to shout us down and say that we must discuss the question at length, but the blokes refused to listen and told them in no uncertain terms where to go.

'We decided to organise a mass meeting of all Cowley Hill men on Sunday morning, and to put pickets on the gates to tell the other shifts going in on Saturday what it was about. Then about 30 of us went round all the other Cowley Hill departments in groups of 6 to tell them what was going on and ask for their support. We didn't put any

THE OUTBREAK OF THE STRIKE

pressure on – we just told them to make up their own minds as to what they did. About a dozen of us got through to the City Road plant, which is in the same perimeter as Cowley Hill and asked for their support. We got a good reception but we left it up to them. The last place I went into was the "old warehouse" where they were all getting ready to come out. One bloke of about fifty or so asked if we were on strike. I said we were, and he said: "About bloody time too. It should have happened 15 or 20 years ago." By about 1.45 am on the Saturday morning everyone was out.

'On Saturday morning at about 5 to 6 o'clock when the 6 am to 2 pm shift was going on, there were groups of pickets standing about 50 yards or so away from the gates who talked to the men as they went in. About five or ten minutes later they had all come out. The 8 o'clock day turn men came, we talked to them and most went straight back home; a few went in and came out straight away saying it was like a morgue inside. The afternoon shift didn't turn up at all because by that time, word had gone right around town.

'At the mass meeting on Sunday morning, there was a crowd of about 2,000. The shop stewards addressed the meeting and recommended that we go back to work. One of them said: "You're a bunch of cowboys. Who's going to give you 2/6d? You're asking for the sky." So, a bloke pipes up from the crowd: "Well, let's ask for the sky then, let's ask for £25 a week" (i.e. 5/-d an hour on the basic pay). And that's how we all eventually came to be out on strike for £25 a week. There was a vote taken and the men voted to stay out and ignore the advice of the stewards to return to work. As the meeting started to break up someone shouted from the crowd: "Let's go to the Sheet Works meeting this afternoon because if we don't tell them the truth, you've no idea what the stewards will tell them." '

The meeting of the Sheet Works, like the one at Cowley Hill, voted in favour of the strike. A third eye-witness gave us the following account of how events developed after the Sheet Works meeting.

'We marched up the road to the Ravenhead–Fibreglass Lodge, 2,000 of us. The rain was sluicing down. When we got there, we stopped outside the gates although a few hot-heads had been saying earlier that we ought to go in and drag them out. Just after we got there, a copper arrived in a Panda car. He got out slowly, took off his flat hat, put on his topper, and came over to us and said: "All right, lads, you just stop where you are and there'll be no trouble." Just imagine it, one copper and 2,000 men – and they all stayed put. The crowd started chanting: "Out, out, out!" Then we sent in one man to tell them that we were out and ask them to join us. A few minutes later one man came out on his own, about 40 a few minutes after that, and a couple of hundred after that. When they were all out, we decided to march on to Triplex, about a mile away.

'We asked the copper which would be the best way to go because we didn't want to cause any traffic congestion. He said: "Don't ask me, I'm only a handrag. You'd better ask the brass." Of course, we didn't have time to wait and any-way it was still sluicing down with rain. So we set off. We'd got three coppers by now – one in a car leading the procession, and one on each side walking. When we got to Triplex we'd already been beaten to it – a load of blokes, crafty buggers, had gone round in a cavalcade of cars so they wouldn't get wet. But it didn't matter because the people at Triplex were already coming out of the gates. Then we decided to march to the City Road plants about $1\frac{1}{2}$ miles away. When we got there, they were just starting their own meeting and came out almost straight away. By about half past three, on Sunday afternoon, everyone was out.'

These eye-witness accounts reveal that at the Sheet Works and Cowley Hill plants where the strike began on Friday, 3rd April, only approximately 350 and 200 men respectively were involved in taking a decision to strike. Other workers in these plants were presented with the situation where a decision to strike had been taken. When the strike spread to the other four factories in St Helens on Sunday,

THE OUTBREAK OF THE STRIKE

5th April, it was known that two plants had already stopped work and that a crowd of 2,000 was marching in support of the strike. The momentum that set the strike in motion was supplied by relatively small groups of employees in relation to the total size of the Pilkington labour force in St Helens. Once the strike was under way, however, considerable social pressure was aroused to induce employees at all the factories to become involved in the dispute. In our investigation 80% claimed that their initial involvement was reluctant. A point that must be borne in mind here is that at the beginning of the strike the situation was confused and no individual worker had any way of knowing what all his colleagues felt about the events that were taking place. That even the aims of the strike were confused is quite apparent. One thing that was clear was that a numerically strong body was supporting the strike. This was the situational pressure that resulted in all the factories being brought to a standstill.

Physical threats and force played no significant role in the outbreak of the strike. The pressure under which people struck was moral pressure. Those people who came out of their factories on the Sunday afternoon did so largely because there were thousands of their fellow-workers outside the gates (though doubtless there were some who were happy to join in): an appeal for solidarity of such weight can hardly be ignored, particularly if it is backed with the prospect of a £10 a week increase in pay. The same pressure, although in a lesser degree, operated on those people in the other factories who voted to walk out, and on all those people – the majority – who had not been at work over the weekend and who did not turn in for work on Monday morning.

Moral pressure is a very potent force. Indeed, it is the potentiality of moral opprobrium which plays a large part in keeping most forms of social organisation together. It is quite as effective in a strike as it is in the outwardly neat and ordered life of semi-detached suburbia. One must also bear in mind that at the beginning of the strike, no one could have known that the dispute was going to last for seven weeks. Few of the strikers questioned in our investigation had believed initially that they were entering into such a prolonged stoppage. Only 16% had thought that the strike would last for longer than a fortnight; the majority had believed that it would be settled in a matter of days. When they first joined in the strike most of the strikers felt that they were giving their support to a dispute that would be quickly settled. They were wrong. But these initial views about the likely duration of the strike must have made people less resistant than they might otherwise have been to the pressures that set the stoppage in motion.

The union and the RFSC

The results of our survey showed that rank and file attitudes underwent significant changes during the course of the dispute. This demonstrates one very obvious fact; a strike is a dynamic phenomenon. During a strike the participants' aims and attitudes do not remain static; they can undergo dramatic shifts in response to the flow of events. Thus, the conflicts that need to be resolved at the end of a strike may be different from the problems that led to its beginning.

One important development in rank and file opinion that occurred during the Pilkington strike was that criticism of and disenchantment with the union grew, whilst sympathy and support moved behind the emergent RFSC.

When the strike began few of our respondents realised that they were embarking on an unofficial strike: 82% of our strikers said that at the beginning they believed the dispute would get union backing. A number of features about the way in which the stoppage began could have created the impression that it would receive union support. In the first forty-eight hours, some shop stewards were advising their men to strike and at a meeting on the fourth day union officials were expressing sympathy with the aims of the strike and saying that it was 'official at branch level'. But the union did not make the strike official. On

Wednesday, 8th April, during the first week of the strike a mass meeting was addressed by Mr David Basnett, a national officer of the GMWU. He informed the meeting that the pay claim should be pursued through the proper negotiating machinery, requested an immediate return to work, and made it clear that the union would not make the strike official.

By the time of our survey in the sixth week, the majority of the strikers we questioned were critical of the way the union had handled the dispute: 79% had formed an unfavourable opinion of the union and 65% were actually in favour of leaving the GMWU. and forming or joining an alternative organisation.

Before the strike some dissatisfaction with the union had already existed. But there had been no attempts actually to leave the GMWU. The union's handling of the strike itself resulted in criticism blossoming forth. Most of the strikers were not active trade unionists. They treated union membership as an insurance policy and as a source of certain services. One thing they regarded their membership as assuring them of was help in the event of 'trouble', and for the Pilkington labour force, the strike that blew up was 'trouble' on an unprecedented scale. Whilst most of the members had not supported the strike, there was a general feeling that since the dispute apparently possessed local support, then it was the duty of the union to take up the aims of its members. It was also regarded as the duty of the union to 'sort out' the trouble that had arisen. But in the early stages of the stoppage, as far as the strikers could see, their union's national officials were doing nothing to resolve the dispute nor to further the aims of the strike. Consequently, union speakers received a very rough ride at mass meetings and the more colourful events of the strike included the picketing of the union's local offices to ask why the dispute had not been made official.

The criticism that it aroused early in the dispute, when the confidence of its members drained away, resulted in the union ceasing to be an effective source of leadership amongst the rank and file in St Helens. Subsequently, the union made a number of attempts to organise support behind its leadership, but was consistently unsuccessful. Early in the strike it organised a ballot to find out whether its members really did support the stoppage. The results of this ballot were never announced, possibly because the union's policy of an immediate return to work was rejected, although the explanation given by the union was that the situation had changed substantially by the time that the results became available. An offer of a £3 increase on gross levels of pay was secured from management after the strike was two weeks old and was used by the union as a bait to secure a return to work. Propaganda was fed to the press from union sources, aimed at discrediting the union's opponents. Eventually a second ballot was organised under the supervision of the local clergy in St Helens in which a majority of 274 out of the 6,246 who voted supported the union's policy of a return to work. But even the successful outcome to this ballot did not enable the union to organise a return to work. Less than 3,000 returned to work during the following week. The union never regained any effective leadership or control of the situation. Indeed, its officials experienced considerable difficulty even in communicating with their members, for at mass meetings their speeches tended to be drowned beneath a barrage of hostile criticism.

At the time when the union was losing the confidence of its members the RFSC emerged into prominence. This committee rapidly became the most effective source of leadership operating amongst the St Helens strikers. The RFSC appeared to be the only body that was responsive to local opinion, the only body that was clearly pledged to further objectives of the strike, and it succeeded in organising impressive demonstrations of support at mass meetings attended by between 3,000 and 5,000 of the strikers. Support for the RFSC amongst the strikers was by no means universal. In our survey a bare majority (54%) expressed favourable opinions about the policies and actions of this committee. But there was considerably more sympathy for the RFSC than for the union and this sympathy was made visible to the public at the mass meetings that were organised by the committee. In contrast, there was no public evidence to indicate that any significant number of the strikers supported the union and its policies. Indeed, the actions of the union at the beginning of the strike had resulted in its grass-roots support in St Helens dwindling to an insignificant level.

The invisible majority

Alongside the developments that occurred in the attitudes of the strikers towards the organisations that were claiming their support, a significant shift also took place in the views of the rank and file towards the strike itself.

Only 19% of our strikers had supported the strike at the outset, but by the time of our survey 53% thought the initial decision to strike had been justified, and 43% were in favour of continuing the strike until the firm improved the terms of £3 on gross rates of pay that had already been offered. Thus there had been a definite movement of opinion in support of the strike. It is entirely possible that support had risen to an earlier peak from which it had since fallen by the time of our survey during the sixth week of the strike. On the other hand it is equally possible that support was gradually gaining ground throughout.

We cannot draw any firm conclusions as to why this movement of opinion should have taken place. The strike began suddenly and it was possibly not until people had thought about the issues and discussed them that they formed a favourable opinion. Perhaps the support that the strike received from the union branch at the beginning, and later from the RFSC resulted in opinion being swayed by the arguments of these groups. It may even have been that attitudes became more favourable after the firm's offer of £3 on the gross pay; the offer showing that strike action could result in tangible gains. All in all there can be no doubt that different individuals were influenced by different incidents and that a combination of factors was responsible for movements of opinion.

Support for the strike became by no means unanimous. Our survey showed a narrow majority (56%) in favour of a return to work on the firm's terms – a balance of opinion subsequently to be confirmed in the 'parsons' poll' a few days later. However, the minority of 43% in favour of continuing the strike was much larger than the 19% who had supported the strike when it began.

The fact that rank and file opinion moved so significantly meant that throughout the duration of the strike the majority were neither consistently for nor consistently against. Our survey shows very clearly that it was impossible to divide all the strikers into opposing camps of supporters and opponents. There were some who had been consistently opposed – 33% and some who had been consistent supporters – 11%. But 56% had at some stage changed their minds – the general movement being in support. That such changes took place demonstrates how misleading it can be to make crude generalisations about the strikers as a whole – for they never were a whole. They were not only divided at particular times over particular issues, but the opinions of individuals also underwent modifications as time went on.

We feel that this changing state of rank and file opinion can help to explain why the strike was so difficult to settle and why it lasted so long. With opinion so evenly balanced it was clearly difficult for the rank and file to be mobilised for a return. The even balance also meant that both union loyalist and the RFSC could find evidence to support their views that their respective policies reflected majority opinion. The union could point to evidence such as the support that was received for its policies at factory gate meetings on 2nd May (even if they were thinly attended), while the RFSC could point to the overwhelming votes to stay out at its regular mass meetings. The consequence of this was that each group could find confirmation of the support that it believed it possessed and that neither group was therefore willing to concede ground.

A second way in which the state of rank and file opinion can help to explain the prolongation of the strike relates to the minimal support the union was able to command. In our survey only 16% held a favourable view of the union's actions during the dispute and 65% wanted to leave the GMWU altogether – this latter despite the fact that 56% favoured the union policy of a return to work. The RFSC on the other hand commanded a much greater volume of support for 54% of our strikers were generally favourably disposed towards its actions and policies – even though a smaller number – 43% – favoured its policy of continuing the strike. Clearly the problem as far as the union was concerned was that most of the people favouring its policy of a return were quite out of sympathy with its organisation and leadership. Only at the end of the seventh week when the RFSC accepted an offer of mediation from the TUC general secretary did the strike come to an end. Previously the strike could not be terminated because no source of respected leadership was available behind which those in favour of calling the strike off could gather in order to become a visible and effective force.

This problem of why the strike was so prolonged since it had so little support at the beginning and possessed only minority support during its closing stages raises the issue of intimidation. During the strike both the firm and the union were convinced that people were being prevented from turning in to work. In our survey we asked those who favoured an immediate return to work why they had not in fact gone back through the factory gates. 63% said it was because they were intimidated, or words to that effect. The remainder explained their continued participation in the strike in a variety of ways, the most frequent of which was the desirability of abiding by majority opinion.

This question of 'intimidation' must be clarified for the fact that many strikers perceived themselves as being intimidated does not prove that they *were* intimidated or that anyone was intending to intimidate them. There were, to be sure, some 'violent' incidents but they were neither numerous nor notably damaging. But the incidents of violence that did occur were given a great deal of press publicity, and the union justified its distribution of hardship payments to the strikers on the grounds that its members had been prevented from returning to work.

What really needs to be explained however is why it should have been those who wished to return to work who felt intimidated – they were after all the majority. The answer is obvious: the majority were either unaware or uncertain of their majority position. No one individual had any way of knowing the thoughts of the remainder of the rank and file. No source of effective leadership was available beneath which those who wanted to end the strike could demonstrate their strength, while those who supported the strike were highly visible at mass meetings and on the picket lines.

The evident support for continuing the stoppage could well have made those who wanted to break the strike fear for their safety. They might also have considered that an attempt to return in the face of such massive displays of solidarity would have involved an open invitation to abuse and other informal social sanctions. Such sanctions were, as a matter of fact, applied following the general return.

Like the pressure that was involved in the outbreak of the strike the intimidation that prolonged it was almost entirely of a moral character. That such intimidation should have been effective can only be explained in terms of the invisibility of that body of rank and file opinion that favoured a return, and this invisibility can only be explained in terms of the absence of a source of leadership sufficiently effective to thrust it into public prominence.

The politics of the rank and file striker

During the dispute suggestions were made from several sources, including union spokesmen, that those who were supporting the strike were inspired by political motives. To assess whether this view had any substance as regards the rank and file strikers we included some questions about politics in our survey.

As would have been expected the majority of the strikers possessed left-wing political sympathies. But their left-wing persuasions were distinctly moderate: 73% were normally Labour Party voters whilst only three out of the 187 strikers who were interviewed expressed support for a party to the left of Labour. Very few of the strikers could be described as political activists; only 13% belonged to a political party (in most cases the Labour Party).

Whilst the strikers did possess political opinions (which of course is perfectly normal) very few saw the strike in which they were involved in terms that could be even remotely described as 'political'. We asked our respondents whether they saw the strike as involving any political or social objectives wider than the domestic issues upon which the dispute was ostensibly based. In most cases this question aroused a vigorous denial that the strike had anything to do with politics. Most of the strikers were clearly offended by the implication that their strike could have any political significance whatsoever. Amongst these respondents who had been consistently opposed to the strike there was a substantial minority (26%) who attributed political motives to those who were supporting the dispute. But amongst the strike's consistent supporters only 16% admitted that their support was inspired in any way by political purposes. There were individual strikers who saw the dispute as an opportunity to launch an attack upon capitalism, to hit at the employing class, to introduce shop-floor democracy into industry, and to further the political education of the working class. Such individuals were, however, a small minority. The dominant view was that work and politics ought to be kept rigidly compartmentalised. It was strongly felt that people ought to leave their political sympathies behind when becoming involved in industrial action. As a generalisation it would therefore be incorrect to attribute political motives to the rank and file supporters of the strike. Furthermore the manner in which the majority repudiated the notion of political significance suggests that efforts by 'subversive' political groups to exert an influence upon shop-floor affairs could stand little chance of success. Individuals with an obvious political axe to grind would automatically be considered suspect.

Representatives of various political organisations who entered St Helens during the course of the dispute in the hope of being able to attract support received little encouragement from the rank and file strikers. Typical attitudes towards such representatives ranged from amusement to resentment at the attempt to encroach, and as we shall see in Chapter Six, when the RFSC is discussed, political factions were regarded in a similar manner at this level.

The experience of the rank and file striker

What was the experience like for the Pilkington employees who were on strike for seven weeks? What did it feel like to go on strike? One thing that is certain is that different individuals reacted to the experience in quite different ways.

To some the strike was an education; it opened their minds; it broadened their horizons; it gave them new insights into themselves and into the society in which they lived. During the dispute some individuals began to think and argue about issues that they had never previously attempted to understand; they discovered abilities within themselves that formerly they never knew they possessed. For these people the strike could be rightly described as a revolutionary experience. The outbreak of the strike to such individuals was not simply a matter of stopping work. Apart from this technical beginning the strike also had a conscious beginning in which they became aware of the radical significance of the step that they had taken. A participant in the walk-out at the Conley Hill plant on the first Friday of the strike described his experience in this way:

'There was a feeling of elation, a feeling of liberation

about the place, even though everyone did seem to be confused about what was going on. The way some of the men were talking it was as though they had done something big for the first time in their lives. And it was the long-service men who were doing all the shouting – the blokes who got up and spoke in the canteen were all long-service men of about forty to fifty years of age.'

To go on strike is to deny the existing distribution of power and authority. The striker ceases to respond to managerial command; he refuses to do his 'work'. A new dimension of living can thus be revealed to the striker; an existence in which 'ordinary' people are able to control events and command the attention of 'them'. The experience of this new reality can transform the striker's perceptions of normal life. What was 'normal' can no longer be regarded as 'natural'. Attitudes towards work and authority become critical as opposed to acquiescent. The sudden acquisition of these insights hits the striker with a blinding clarity; hence his feelings of liberation and elation.

From these initial revolutionary sensations the development of the strike, for some individuals, offered an interesting and even an exciting education. Those who supported the strike and who were actively involved in its organisation found themselves administering affairs and attracting public attention on a scale that they had never previously experienced. Some individuals enjoyed the strike; it was much better than being at work.

As one striker told us, 'I enjoyed it without a doubt. I enjoyed the freedom it gave me to express myself which was something entirely new to me after working in factories. It gave me a little bit of scope to use the abilities that I have. Educationally it was great. I liked the companionship of the pickets; I liked the feeling of being involved in a struggle; there was a great feeling of "oneness" that I really enjoyed.'

This experience of the strike, however, could have been shared by no more than 900 of the strikers. Our survey results suggested that no more than this proportion of those involved were consistently in support of the strike and therefore in a position to react to the dispute in this manner. There were other employees who were morally shocked and repulsed by the outbreak of the strike. To some of Pilkingtons' older, long-service employees the dispute shattered their traditional conceptions of the nature of the social order within which they were living. For most of the employees, however, strikes were sufficiently common features of the national industrial scene to have lost their capacity to produce reactions of moral outrage. To the majority who felt no personal desire to strike the outbreak of the dispute was simply annoying: it impeded their enjoyment of the normal routine of life.

Those who experienced no sense of elation and liberation at the start of the dispute and who were not personally involved in the conduct of strike activities found the prolonged stoppage a troublesome experience. They were thrust into a limbo from which there was no obvious way out. Throughout the strike they lacked a full knowledge of the events that were taking place for they did not have access to all the information that was available to those playing more central roles: their main sources of information must have been press and TV reports interlarded with street-corner gossip. If all the strikers received letters from the company, while some attended mass meetings and read leaflets distributed by the union and the RFSC, how were they to know which source was providing accurate information? They were not.

On some key issues a rank and file striker's knowledge could amount to no more than guesswork. He could not know whether or not the union had been aggressive in its negotiations – he could not even be certain of the significance of Pilkingtons' pay offer. He knew that the £3 offer was not to be added to his basic pay (the basis of calculation for subsequent bonuses), but did this mean that it would be a bonus that some workers might not be able to earn? Some certainly placed this interpretation upon it.

Apart from the feelings of confusion and uncertainty that a lack of reliable information inevitably generates, a further troublesome feature of the striker's predicament was the hardship that he had to face.

The severity of hardship varied according to whether other members of the strikers' households were working and whether other members of their families were eligible for social security benefits. But some hardship was involved for everyone. Standards of living had to be stripped to the bare essentials; many were obliged to go into arrears on rent and HP payments; savings were swallowed up; in some households durable goods had to be sold; some strikers were forced to receive charity from relatives and friends. In addition to all this was the sheer tedium of being at home all day: with little money available any holiday spirit that may have been around at the beginning quickly evaporated.

For the many whose experiences of the strike were of this nature there were no feelings of liberation and elation. To quote one participant, the impact of the dispute had been 'to make it likely that it will be another hundred years before a strike occurs at Pilkingtons again.'

The strike affected its various participants in radically different ways. This is one reason why it is misleading to attempt to generalise about the rank and file as a whole. The only common element in their experiences of the strike was that for everyone it was an abnormal sensation. Everyone learnt something from the strike, but the content of the lesson varied.

The influence of the rank and file

It was stated at the beginning of this chapter that the rank and file participants play a decisive role in any dispute. It is a truism that only by rank and file action can a strike begin or end. But what this analysis of the rank and file during the Pilkington dispute has revealed is that their behaviour is not necessarily a direct response to majority attitudes and opinions. The strike began with only minority support amongst the rank and file and it was prolonged for seven weeks despite the wishes of a majority of the strikers to have it called off earlier.

For rank and file opinion to lead to decisive action it must be mobilised into an effective force; rank and file opinion must become visible and achieve some degree of social organisation. The mobilising agent may be a spontaneous movement (as when the strike began) or a leadership group (as the union and the RFSC were attempting to be throughout the duration of the dispute). Leadership groups themselves constantly respond to the indications that they receive regarding the state of rank and file opinion. Mass opinion is involved in a continuous and reciprocal process of interplay with the policies and actions of leadership groups. In a strike the rank and file can never become mere puppets. But there are circumstances in which large sections of rank and file opinion can be rendered impotent. The absence of an appropriate leadership group can prevent a dormant body of opinion being mobilised as an effective force. This happened during the Pilkington dispute amongst those who favoured a return to work but who were not mobilised to become a visible and effective group.

It is quite conceivable that in other industrial situations groups of workers who are prepared to strike will be restrained because the situation inhibits the revelation of their military outlooks. Similarly circumstances can easily be envisaged in which a strike could be called off in the face of immobilised opposition from rank and file opinion.

To understand the role of the rank and file in any strike it is clearly necessary to examine its relationship to other elements in the situation, particularly those groups who are attempting to play leadership roles. Only when the rank and file is examined in this broader context is it possible to explain why and when the decisive actions required to set in motion or to conclude a strike are taken.

Chapter Four

THE UNION

During the strike the GMWU was under attack from all sides. Rank and file members believed that the union had let them down by failing to pursue the objectives of the strike as strenuously as possible. The RFSC was even more critical of the union's role in the dispute, alleging that the union had been a tool of the bosses, placing the interests of the firm before the wishes of its members. It was also alleged that the union had been controlled undemocratically, and that it had failed to reflect and respond to the interests of its ordinary members. Nor did the GMWU escape criticism from the firm. The company claimed, at the Court of Inquiry held during the dispute, that the union was at least partly to blame for the strike for failing to co-operate earlier in reforming the company's wage and negotiating structures in ways that the firm felt could have prevented the strike from occurring.

The GMWU accepted none of these criticisms. Union spokesmen were quite prepared to admit that there had been inadequacies in the union's organisation, but they claimed that these weaknesses had been acknowledged by the union and were in the process of being rectified before the strike. Their spokesmen were also willing to admit, with hindsight, that their organisation had made a number of tactical errors during the dispute. But they were adamant in their conviction that the union's policies during the dispute had been fundamentally right, that the union had been controlled in a thoroughly democratic manner, and that it had represented the interests of its members as effectively as possible under difficult circumstances.

Before we can disentangle the truth about the role played by the union during the strike it will be necessary to outline how the GMWU in St Helens was organised prior to the dispute.

91 Branch

Production workers at all the six Pilkington factories in St Helens were organised by the GMWU. In 1964 the company had entered into a condition of employment agreement under which all new employees were obliged to join the union, and as part of their working relationship the firm assisted the union by deducting members' subscriptions from their pay packets. At the time of the strike there were approximately 7,400 GMWU members in the six St Helens factories. These members were organised in the '91 branch' of the union. Its 7,400 members made this one of the union's largest branches in the country. In financial terms its members' subscriptions resulted in an income to the union of approximately £40,000 per year. The branch maintained its own committee rooms in St Helens and a whole-time secretary.

To keep all the 7,400 members involved in the affairs of the branch presented difficulties. Within the union it was widely felt that the branch had grown too large and that it needed to be subdivided. The single branch secretary did not have enough time to attend to all the problems that arose in the factories in which his members worked. The need for a reorganisation of the union, possibly into branches based upon each factory, in order to make greater contact possible between branch officials and shop-floor members was generally recognised within the union.

With such a large branch it would have been difficult to hold meetings that could be attended by the entire membership, and in fact no attempts to arrange such meetings were made. Only shop stewards were allowed to participate fully in the monthly 'branch' meetings. Whether these events should be described as branch meetings is really debatable. The GMWU's rule book makes no allowance for participation in branch affairs being restricted to shop stewards, and after the strike union officials acknowledged that the branch had been run unconstitutionally. Other members could attend branch meetings but they were not

BRANCH

allowed to vote upon decisions that were taken. In the six Pilkington factories there were 104 shop stewards active at the time of the strike, all of them elected from the shop floor. It was a shop steward's job to act as the representative of his men on any issues that needed to be raised locally with management, and also to represent his men at branch meetings where generally policy matters could be discussed and voted upon.

In a formal sense the branch was run in accordance with principles of representative democracy, but in practice problems arose regarding the representation of the views of rank and file members. Firstly, there was the problem that rank and file members were unable to participate directly in policy decisions at any level in their union. Secondly, there was the problem that the shop stewards who were supposed to represent their members at the branch often did not bother to attend the branch meetings. At a typical branch meeting there would rarely be more than twenty shop stewards present. This meant that the majority of the union's members did not have any open channels of communication with their branch. As a result grievances that arose on the shop-floor would not necessarily be raised at branch meetings where they could either be directly acted upon or passed upwards for consideration at the higher negotiating levels. Similarly information about policy and action at branch level and above would not necessarily filter down to the rank and file. Since most shop stewards did not attend branch meetings the affairs of the branch tended to be managed by a small number of activists in conjunction with the union's full-time officials. At the monthly branch meetings there would normally be only between fifteen and twenty shop stewards present and a great deal of the branch's business had been delegated to an even smaller committee which was set up approximately eighteen months before the strike. This sub-committee had not been subsequently re-elected and did not have a rigidly determined membership. According to one of the active shop stewards the practice was for the branch chairman to co-opt anyone who displayed sufficient interest to serve on this committee. The effect of this working arrangement was to place effective control of the affairs of the branch in the hands of a very small number of activists.

For the purpose of obtaining a view of the strike from the union's point of view we interviewed six of the active shop stewards; one from each of the St Helens factories that were involved in the strike. Each of these men had been amongst the few who played a central role in the branch prior to and during the strike. Five out of the six were also union representatives on the JIC. They were represented on all the union committees that played a role of any importance during the strike. These stewards were consequently excellent informants from the point of view of obtaining an account of how the union had acted during the period leading up to and during the stoppage. Apart from being close to the centres of activity within their union the shop stewards that we interviewed had also been loyal to their organisation throughout the strike. They had been personally involved in key decisions that the union had taken and had consistently supported these decisions throughout the dispute. Initially, these stewards were nominated to us by the union's regional office as suitable individuals to approach to gain an account of the strike from the union's point of view.

A trade union such as the GMWU is necessarily a complex organisation containing a number of inter-related but distinct decision-taking levels. During the strike union policies and actions were initiated from three different levels of authority: the branch, the region, and the head office. One of the union's problems during the dispute, as we shall see presently, was that on a number of occasions decisions that were taken at different levels were not fully co-ordinated. In view of the existence of these distinct levels of decision-making it is slightly unrealistic to attempt to construct a single account of the strike from the union's point of view. Our informants were most centrally involved in branch affairs and therefore the account of the strike that is presented in this chapter deals basically with the dispute

BRANCH

from the point of view of the union's branch. Our informants were involved in decisions taken at regional and head office levels during the strike, but in a marginal capacity in contrast to their degree of involvement in branch decisions. Because our informants were personally involved in most of the key decisions taken by the union their accounts of the strike do constitute a satisfactory basis for understanding the policies that the union followed. But it must be pointed out that some individuals who were involved in the strike on the union's side, especially the regional and head office officials, possessed opinions and reacted to events in a rather different manner than this chapter will suggest.

In addition to the shop stewards we also spoke to a number of the full-time officials of the GMWU who had been concerned with the strike, mainly to seek clarification on questions of fact. We also were able to refer to the written evidence that the union placed before the Court of Inquiry. However, the view of the strike that is presented in this chapter will be essentially based upon the accounts given by the shop stewards, who will subsequently be referred to as the 'union spokesmen'. They were all interviewed during the week immediately following the end of the strike. The interviews were informal; each respondent being invited to describe the outbreak and development of the strike as he had seen it, and to explain the various actions that the union had taken. Needless to say no two of the shop stewards told identical stories, but there were sufficient points in common between their accounts to enable a general union view of the strike to be constructed.

Six shop stewards may appear to be a rather small number of individuals to have interviewed in order to construct a picture of the strike from the perspective of the union. The fact was, however, that the number of individuals who had played an active part in union affairs before and throughout the whole of the strike was very small. This will become clear in the account that follows.

The union spokesmen regarded the active few who were involved in branch affairs as a small minority who were interested in and committed to their union's cause. The major problem that they believed they faced was the apathy that the rest of the members displayed towards the union. They in no way regarded the situation as being one in which the branch's structure was thwarting a widespread desire for greater participation. As far as they could see the mass of the members just did not want to be bothered with union business.

The general state of apathy that gripped the membership was seen to include the bulk of the shop stewards, most of whom could not summon sufficient interest in union affairs even to attend the monthly branch meetings. There was rarely any difficulty in persuading members to offer themselves for nomination to the position of shop steward, and nominees were often elected unopposed on a show of hands at shop-floor meetings. The problem was that once they had been elected few of the stewards bothered to involve themselves in branch activities.

Before the strike the reorganisation of the branch had not been treated as a matter of urgency because there was no general pressure for such a reform. Members and shop stewards were not demanding greater opportunities to participate. According to our spokesmen those shop stewards who were active were distressed at the way in which they were stranded in a sea of apathy. Reform was obviously required but there was no reason to believe that it needed to be dealt with as an urgent issue. Once the strike broke out the weaknesses that existed in the branch's organisation became immediately evident. The absence of effective channels of communication made it difficult to assess and control the situation. With the mass of its members relaxing in a state of apathy, however, the branch saw no signs that the outbreak of a strike was imminent.

The Joint Industrial Council

Negotiations between the GMWU and Pilkingtons were conducted primarily through the JIC. The GMWU was

THE JOINT INDUSTRIAL COUNCIL

the only union represented on the employees' side of this body. Factories in the Pilkington group in which the labour forces were not organised by this union, and grades of employees who were not represented by the GMWU were catered for by quite different negotiating institutions. The union members in most of the factories had one representative on the JIC. Of the twenty-two union representatives on the council fifteen were directly elected from the shop-floor, the remaining seven being full-time union officers. The preponderance of lay representatives on its side of the negotiating table was a feature in which the union spokesmen displayed considerable pride. They felt that the composition of this top negotiating body highlighted the democratic nature of their union. Negotiations of the highest level were conducted by men who were themselves workers on the factory floor and who were directly elected by their shop-floor constituents.

Before the strike, however, the union had already openly expressed concern about the existence of weaknesses in the JIC machinery. One problem was that the members' apathy towards union affairs in general resulted in little interest being taken in the election of shop-floor representatives to the JIC. It was by no means unusual to find only twenty or thirty members voting in elections for representatives who were supposed to represent 1,000 or more employees. Consequently the machinery of elections was failing to ensure that the union representatives on the JIC reflected shop-floor opinion. Already before the strike the union had agreed with the firm that the JIC machinery was in need of an overhaul.

Fundamentally the union spokesmen believed that the JIC was a good institution, and that the way in which it enabled shop-floor employees to be involved in top level negotiations was highly desirable. It was also felt that the record of the JIC was proof of its value from the union's point of view. The earnings of Pilkington employees who were covered by the Council were above the average levels of earnings for manual workers throughout industry as a whole, proving that the union had been able to use the council to further the interests of its members in an effective manner. The union felt that criticisms which claimed that both the JIC and the union had been used by the firm to protect its own interests were completely without foundation. The record proved that the JIC had been an instrument which the union had been able to use effectively to further the interests of its members.

However, the union did feel that its negotiating procedures with the company were in need of modification. As the spokesmen saw the situation the gradual growth in the size of Pilkingtons had led to the JIC becoming too detached from the shop-floor and reforms were required which would involve employees at factory floor level more fully in negotiations. Negotiating institutions to complement the existing JIC, the union spokesmen believed, were needed in every factory. This would have the effect of drawing shop stewards and their rank and file members more closely into negotiating processes. In addition, the union believed, such reforms would make it easier to rationalise the firm's wages structure which had become extremely complex.

Agreements reached at JIC level had to be implemented in numerous factories and applied to many different types of work, with the result that anomalies had crept into the company's pay structure. Individuals doing very similar types of work could end up earning widely differing amounts of money. Some jobs which at the factory-floor level were felt to deserve extraordinary financial compensation could not be adquately paid under the terms of JIC agreements. At workshop level the JIC system allowed hardly any room for negotiations at all, for local works managers and shop stewards were not allowed to step beyond the terms of JIC agreements. This meant that every new pay agreement negotiated through the JIC resulted in considerable wrangling over the interpretation of the agreement at factory level, the outcome of which was to create anomalies in the pay structure that satisfied no one. The union spokesmen wanted to preserve the JIC as a body for laying down general guidelines, but wanted industrial relations at the factory level to be institutionalised in a way that would enable the manner in which JIC agreements were implemented to reflect the varying needs of different factories and different types of work.

The union spokesmen were aware that weaknesses existed in the negotiating procedures within the company and recognised that the responsibility for changing these procedures was partly the union's, but as they saw the situation it was mainly the minority of activists who were conscious of these weaknesses. At the factory level most members appeared apathetic although the union had warned the company before the strike that weaknesses in its system of industrial relations and the anomalies in the pay structure to which they had given rise could lead to future trouble and unrest.

It is worth pointing out that the problems facing the GMWU in St Helens were not peculiar to this trade union. Problems of involving their members more fully and of finding a place for workshop negotiations in systems of industrial relations are issues that are faced by most trade unions in contemporary society. Many unions, indeed, are in the process of re-evaluating their own structures and negotiating procedures in order to try to find solutions to these problems. The difficulties that the 91 branch of the GMWU in St Helens faced before the eventual strike were quite typical of the problems that confront the contemporary trade union movement as a whole.

The calm before the storm

Whilst they would not have described themselves as complacent in the period preceding the dispute the union spokesmen saw no reason to believe that a strike was about to break out. They were aware that anomalies had crept into the pay structure and were provoking discontent amongst some sections of Pilkingtons' labour force, and

THE UNION

they were aware that deficiencies in the union's branch organisation and negotiating procedures were preventing much of this discontent from being brought out into the open and dealt with. But they believed that the union would be able to carry through the necessary reforms before this discontent could develop into an explosive situation.

To the union spokesmen discontent appeared to be limited to a few departments, in particular factories where the employees had been badly affected by the anomalies in the company's pay structure. In such sections of the labour force they were aware that grievances were rife and that labour-management relations on the shop-floor were strained. As far as the spokesmen could tell, however, such groups constituted only a minority of the union's rank and file membership. Most of the employees in the firm appeared to be satisfied with their earnings which compared favourably with levels of pay that could be obtained elsewhere. Although problems clearly existed in the firm the spokesmen did not believe that the union was facing an explosive situation.

Their reactions when the strike suddenly flared up were therefore characterised by surprise. One of the shop stewards who was interviewed explained that even after it had all actually happened he still could not understand why the strike had occurred. As far as he knew most of the workers had been satisfied with their wages and conditions of work. That such men should suddenly have become involved in a strike was totally inexplicable. Another shop steward summed up the prevailing feeling amongst the spokesmen when he claimed that the form in which the strike had broken out had completely surprised him. He had believed that the atmosphere in the firm had been generally calm. There were known trouble spots where groups of workers had special grievances and in time such trouble might have spread through the entire labour force. But steps were being taken to sort out the trouble and the necessary reforms were expected to take effect before an explosive atmosphere spread throughout the firm. The bulk of the

AN EQUIVOCAL BEGINNING

labour force was seen as slumbering in a calm state of apathy.

Consequently when the dispute did break out the branch was taken by surprise and no plans had been formulated to deal with such a situation. Perhaps the branch could be excused of lacking foresight. However, as we have already seen, very few of the rank and file workers in the firm had been expecting the strike. Also the union's organisation had been under no growing and sustained pressure to deal with grievances that were unsettling the labour force. It would really have been surprising if the branch had been expecting a strike and if it had made itself fully prepared to deal with such a contingency.

An equivocal beginning

When the strike broke out the branch's handling of the situation was characterised by a complete lack of confidence that can only be explained in terms of the extent to which the strike took the union by surprise and in terms of the local branch's lack of experience in dealing with such a situation, for no similar strike had occurred at Pilkingtons within living memory.

The strike began in the Flat Drawn shop at the Sheet Works factory. This particular shop was a known centre of discontent and the union's senior shop steward in the factory was not surprised when at 12.30 pm on Friday, 3rd April, the management requested his assistance in sorting out a dispute that had arisen over the bonus payments to a group of workers in this department. This particular steward was amongst the union spokesmen who were interviewed and according to his own account of events he arrived in the Flat Drawn department at 1.30 pm, prepared to sort out what appeared to be a spot of localised trouble. Upon reaching the department, however, he was informed by the men that it was not the bonus question that was really bothering them: what they wanted was an increase of 2/6d per hour on their rates of pay. This demand for an extra 2/6d per hour had previously been brought before a union branch meeting where it had been accepted for processing after a current pay agreement with the company had been fully implemented. The shop steward's advice to the disgruntled workers who had downed tools was, accordingly, that what they were demanding was a JIC matter that could only be dealt with in due course through the proper channels. This steward's advice was ignored and a group of workers from the Flat Drawn section proceeded to other sections of the factory soliciting support for their claim for 2/6d per hour and encouraging the rest of the employees to join in the dispute. In this factory the strike clearly began in opposition to union advice and the recommendations of union officials were subsequently shouted down at works meetings.

Following the outbreak of the dispute at the Sheet Works a group of the strikers went to a second Pilkington factory at Cowley Hill, succeeded in organising a works meeting, and managed to obtain support for both their strike and their pay demand. So by the Friday evening two of the Pilkington factories had become involved in the strike which at that stage was clearly against the advice that had been given by union officials.

On the Saturday a union branch meeting was held and attended by the unprecedented number of eighty shop stewards. Some of the regular activists were surprised to find that there existed so many shop stewards in St Helens who were eligible to attend the branch meetings. This meeting agreed that the strike was totally in violation of agreed procedures and recommended an immediate return to work. There was no intimation whatsoever that the strike had union support.

By the next day, however, many of the branch's officers had undergone what could only be described as a crisis of confidence. Groups of strikers from the two factories that were already involved in the dispute went to the remaining Pilkington factories in St Helens. The shop stewards and the union members in these factories were confronted by crowds of strikers shouting, 'All out, £5 now,' and the shop

AN EQUIVOCAL BEGINNING

stewards whom we interviewed who faced this situation were uncertain as to how they should handle it. In view of the decision of the branch meeting that had been held on the previous day, the orthodox course for the shop stewards to have taken would have been to encourage the men in their factories to remain at work. What happened, however, was that senior shop stewards in each of the four factories involved explained the situation to their men and either offered no advice at all or suggested that the men should acquiesce in the pressure to which they were being subjected and join in the strike in order to avoid unnecessary conflict. The effect of this equivocation on the part of some of the shop stewards who were known to be active in branch affairs was that many of the men who became involved in the strike at this stage believed that they were entering the dispute in accordance with union instructions. Many shop stewards who behaved in this way were not, in fact, supporting the strike. They were acting in order to prevent an ugly situation from arising. The impression created in the minds of many rank and file employees was, nevertheless, that the union was not opposed to the dispute.

Whether the strike could have been brought to a rapid conclusion if the union had taken an absolutely firm stand during the first weekend of the stoppage cannot be ascertained. What is beyond doubt is that the equivocation of many senior shop stewards during the outbreak of the strike led to considerable confusion amongst the rank and file as to whether or not the strike had been started with the support of the union. This initial confusion was to contribute to the union eventually losing complete control of the strike situation.

By Monday, 6th April, all the six Pilkington factories in St Helens were involved in the strike and a further union branch meeting was held. This meeting was attended by an even larger number of shop stewards than had been present on the previous Saturday. At its Monday meeting the branch astoundingly reversed the policy towards the strike that it had adopted only two days earlier. Whereas on the Saturday the branch had recorded without dissent a vote in favour of an immediate return to work, on the Monday the branch decided again without dissent to support the strike. It was agreed to make the strike 'official at branch level'. The branch's position appeared to have been completely reversed. However, according to the union spokesmen, what had happened over the weekend was not that the attitudes towards the strike of the individuals who were responsible for the branch's decisions had changed in any significant way. On the Monday the shop stewards who were interviewed were without exception personally still in favour of an immediate return to work. They explained, however, that they were faced with a situation in which the strike had by then gained support in all of the six Pilkington factories in St Helens. Under these circumstances they felt that it would be unwise for the branch meeting publicly to condemn the strikers. By appearing to be in sympathy with a strike the outbreak of which they had disapproved they hoped that the branch would be able to take control of the situation. Also they felt that although strikes in contravention of normal procedures were to be deplored, in view of the fact that a stoppage was already under way there could be no harm in taking advantage of the situation to bring pressure upon the firm to offer an increase in rates of pay. It was for these reasons, according to our union spokesmen, that the branch decided to go on record in support of the strike, the objective of which was declared to be a £10 increase on basic rates of weekly pay with £5 being the minimum amount acceptable as an interim settlement.

To the rank and file strikers the branch's subtle reasons for supporting the strike could not have been apparent. Many strikers formed the impression that their union branch had given the strike its unreserved backing. The rank and file could not know that the resolution that had been passed at the branch meeting in support of the strike had been supported by people at least some of whom still disapproved of the manner in which the strike had broken out. Another subtle distinction the significance of which could not have

AN EQUIVOCAL BEGINNING

been apparent to many of the rank and file strikers was contained in the phrase 'official at branch level'. In the GMWU strikes involving as many members as were involved in the Pilkington dispute can only be declared official by the national executive committee. The 91 branch simply did not have the power to declare the dispute at Pilkington's official. Therefore the form of words used by the branch in expressing its attitude towards the strike was really meaningless. Nevertheless, in the minds of many rank and file strikers the impression was created that the dispute was official and that they would begin to receive strike pay in due course.

Although it was not evident at the time the reactions of the branch during the outbreak of the strike were to create unforeseen difficulties for the union. The branch's subtle and equivocal reactions had resulted in the appearance being given to many members that the union was initiating the strike. This partially false impression that was created was symptomatic of the inadequacy of the state of communications between the branch and its members. Faced with a crisis situation the branch proved to be incapable of communicating to the rank and file just what its attitude towards the strike was. From this point until the end of the strike the union was unable to give its rank and file members a clear picture of what its policies were and what actions it was taking.

Yet according to the union spokesmen at this early stage in the strike the branch believed that it had grasped control of the situation. Strikes committees were set up at each factory and a central co-ordinating committee was instituted by the branch. It was believed that the branch had succeeded in handling the situation competently.

The equivocal manner in which the branch reacted during the outbreak of the strike may appear to be strange. One must bear in mind, however, that at this stage in the dispute the reactions of the 91 branch were being determined by a group of shop stewards who had been taken by surprise and who had no experience of dealing with the type of situation that had arisen. The shop stewards whom we interviewed declared themselves to have been totally uncertain about the proper course of action to take. Throughout the first weekend of the strike in this state of uncertainty they responded to events in what they judged to be the most sensible manner.

Mass humiliation

On Wednesday, 8th April Mr David Basnett, the union's national industrial officer responsible for the glass industry, visited St Helens and addressed a mass meeting of the strikers. The message that he delivered to the meeting was that as far as the union was concerned the strike was in breach of established procedures and was totally unofficial. The strikers should immediately return to work so that negotiations on their demands, which the union was quite willing to take up, could take place in a calm atmosphere. This message aroused the anger of the crowd and Basnett was given a hostile reception. Both the message and the same reception were repeated at a further mass meeting during the following week. As far as the rank and file strikers could see a complete rift had arisen between their local branch and the union's national hierarchy. The local branch had apparently given the strike its unqualified blessing, whilst a spokesman from the union's national headquarters was completely repudiating the strike. It was these events that led the rank and file to feel that they had been betrayed and let down by their union.

At this stage in the dispute differences of opinion had arisen between the union's local branch officials and the national hierarchy. The branch wanted the strike to be declared official and continued to urge this course of action throughout the first weeks of the stoppage. At a national level the union was unwilling to adopt such a course of action. It was this situation of the union apparently refusing to support its members and their branch that led to criticism of the union mounting. In fact, however, the rift that had arisen between the branch and the national hierarchy was much narrower than most of the rank and file realised. The policy that the head office was advocating had been agreed to by the whole of the union's side of the JIC on which a number of branch officials were sitting.

The local branch was not really completely behind the strike. Although the branch had voted unanimously to support the dispute at least some of the participants at that meeting had been voting in support of a strike of which they really disapproved largely as a strategy for gaining control of the situation and because they believed that since the strike had begun the situation should be exploited to bring pressure to bear upon the company. At a national level the union's policy, which was being pursued by David Basnett, was virtually identical. From the outbreak of the strike Basnett had begun to use the situation to bring pressure upon the company to concede the pay increase that the strikers were demanding. However, he was unable to express public approval of a strike that was so clearly in violation of procedures to which his union was committed. He was bound to request the strikers to return to work. Those individuals who were playing a central role in the union's affairs understood Basnett's position. Needless to say the rank and file did not. The ordinary strikers had no idea that during the first weeks of the strike Basnett was attempting to press the strikers' demands upon the firm at the negotiating table.

The spokesmen whom we interviewed felt that Basnett's handling of the mass meetings had been rather tactless. Following the declaration of support for the strike by the local branch the membership had been expecting to hear similar words from Basnett. When their expectations were rudely flouted Basnett instantly became a target for hostility. The spokesmen felt that the mass meetings could have been handled more skilfully. Whilst it was appreciated that Basnett could not bestow official blessing upon the strike it was felt that he need not have repudiated it so rigidly. Could he not have created the impression that he was really on the strikers' side without giving the dispute his official support? The impression that the spokesmen formed was that Basnett was not at his most effective when addressing mass meetings. Also he had only recently been given responsibility in the union for the glass industry, and consequently he was unknown to the crowd and possessed no personal reputation upon which he could base an appeal. Once he appeared to be dissociating himself and the union's national hierarchy from the strike he immediately lost all hope of gaining the rank and file's respect and subsequently he had difficulty even in obtaining a hearing in his attempts to address further meetings. Some of the union's full-time branch officials believed that the branch had behaved tactlessly by giving indications of official support during the outbreak of the strike. If the branch had maintained the position that its stewards had adopted on the Friday and the Saturday, or if Basnett had succeeded in creating the impression that he sympathised with his members and their demands, the union might have retained a position of leadership in the situation. As it was control of the strike slithered completely away from the union's grasp.

The spokesmen felt that their union's position had been misunderstood. Once the strike was in progress the union was unable to communicate the true nature of its position to the rank and file strikers. The irony of the situation was that the union was acting largely in accordance with the wishes of the rank and file. It was using a strike that neither the union nor the rank and file had initially sought to secure whatever benefits could be obtained for its Pilkington employees. In acting in this way the union's local and national officers were basically united. Unfortunately the union had conveyed the impression that a wide rift had occurred between the branch and the national hierarchy and that the latter was unwilling to take any steps in support of the objectives of the strike. This impression was created by the steps that the local branch and the responsible national officers had taken during the early stages of the dispute. Each step in itself had seemed a sensible tactic to the individuals who took it, but collectively the series of steps taken by the union had completely unintended and, from the union's point of view, disastrous consequences.

By the end of the second week of the strike local support had drained away from the union. The branch's strike committees had become sterile, and an unofficial committee had been formed claiming that only it could speak on behalf of the rank and file strikers. This was a claim to which the majority of the rank and file were by then willing to give their assent.

The £3 offer

As soon as the strike began the union representatives requested a meeting of the JIC. With the strike under way the union side of the council, on which five of our informants were sitting, took up the demand of the strike for an immediate £5 increase on basic rates of pay. This pay demand was justified by the union representatives on grounds of comparability with other industries in which rates of pay had recently been increased. Initially management refused to negotiate until there had been a return to work. David Basnett, leading the union team, persistently argued that he would be unable to secure a resumption of work until a favourable offer had been received. Eventually the management was induced to sit at the negotiating table. The bargaining then proceeded with Basnett presenting the union's case in a manner that our spokesmen who were on the committee described as brilliant. Whatever doubts they possessed about Basnett's competence as a public speaker, at the negotiating table their estimation of his ability was first class. Management was finally induced to offer an increase of £3 on gross rates of pay with a promise of a complete review of the company's wage and negotiating systems to follow. Officials from the Department of Employment and Productivity who were present expressed the opinion that this was a fair offer. The union's side of the JIC was convinced that no amount of further pressure would result in management raising its terms, and therefore decided to recommend an acceptance of the increase. The £3 that the firm offered was less than the union had originally asked for but the feeling on the union side was that by any standards it was a good pay increase to be able to offer the members particularly as a complete review of the company's wage structure was to follow. It was agreed to apply the £3 as a flat increase all round rather than to apply the increase on to the basic rates of pay, because both the management and the union sides of the JIC were convinced that anomalies in the company's wage structure constituted the major problem that they would have to tackle in the future. To have applied an immediate increase to basic rates of pay would have had a multiplier effect upon bonus and shift premiums thus widening existing anomalies. In view of the forthcoming overall review of the company's pay structure it was considered that to widen existing anomalies would be unwise, and therefore it was agreed to apply the money offered as a flat increase all round.

This agreement was concluded in the JIC on 20th April, after the strike had lasted for just over two weeks. Following this offer the Pilkington factories outside St Helens that had become involved in the strike returned to work. In St Helens, however, the £3 offer was to receive a different reception.

Despite the damage that had been inflicted upon their union's reputation at the earlier mass meetings the union spokesmen felt that the pay offer which it had negotiated would redeem the situation. Once their success at the negotiating table became known they believed that the image of their organisation would once again become creditable and that an immediate return to work would follow.

On 21st April the union's JIC representatives placed the terms that had been agreed in the Council before a branch meeting of shop stewards in St Helens. Several members of the RFSC that had been formed during the previous week were shop stewards and were present at this meeting. According to the accounts of the union spokesmen this

THE £3 OFFER

branch meeting was unanimously in favour of accepting the terms that had been negotiated as a basis for a return to work. They recalled that everyone at the meeting was agreed that Basnett and his colleagues on the JIC had done a tremendously good job. Having received such a reception at the branch the union spokesmen believed that they were home and dry.

However, it was decided by the branch to refer the agreement to a mass meeting of the rank and file for their approval. The union's JIC representatives expressed no strong objections to this proposal since they did not doubt that their agreement would be favourably received. That same afternoon the RFSC had been planning to hold a mass meeting of its own, and it was decided to turn this into an official union meeting at which the union's officers could present their recommendations.

Events were to prove that in agreeing to submit itself to a mass meeting in this way the branch had made yet another tactical blunder. According to the union spokesmen a number of things went wrong at the mass meeting. Firstly, the initial sight of David Basnett so incensed the crowd that he had difficulty in outlining the terms that had been won at the negotiating table. Secondly, the poor quality of the loudspeaker system increased Basnett's difficulties. Thirdly, there appeared to be a hard core of militants in the crowd who had come prepared to shout down any terms that the union had to offer. Fourthly, the RFSC members who had been present at the branch meeting failed to communicate this meeting's recommendation that the terms should be accepted and instead appeared to convey the impression that the agreement was a poor one which should be rejected.

Members of the RFSC had quite different recollections of the events that took place at the branch and mass meetings. Those who had been present at the branch meeting of stewards were convinced that the meeting had *not* agreed to recommend the offer; what was decided was simply to refer the agreement to a mass meeting. Also the RFSC members denied that they had attempted to prejudice the mass meeting against the offer. They claimed that their committee simply accepted the decision of the strikers.

However, the union spokesmen felt that they had been 'sold down the river' and outwitted by a treacherous opposition. Once again the branch had chosen to adopt tactics which produced just the opposite results to what had been expected. As a result of its own misconceived tactics combined with the more successful manoeuvring of its opponents the union was left with its hard won agreement having been rejected out of hand, without having regained any sympathy and support from the rank and file, and without there being the slightest prospect of securing more favourable terms from the firm with which to lead a return to work behind the union banner.

The opposing forces

Throughout the remaining five weeks of the strike the union spokesmen were firmly convinced that most of their members wanted to return to work on the terms that had been negotiated. They were also convinced that the policies which the union had pursued had been basically correct and in accordance with the wishes of its rank and file members. The difficulty, as the spokesmen saw the situation, was that the excellent nature of the terms that had been negotiated had not been effectively communicated to the membership, and that the members did not properly understand what the union's policies and actions had been. They believed that the union's difficulties were essentially questions of communication. With its remaining active supporters in 91 branch possessing these convictions the union did not deviate from the policies that it had adopted but embarked upon a series of attempts to publicise its policies in order to regain control.

A branch committee of shop stewards who had remained loyal to the union, consisting mainly of the long-standing activists and JIC members, was set up to organise support for the union's policies. All our spokesmen were members of this committee. Leaflets were distributed in places such as Labour clubs and attempts were made to use the press in order to reach the rank and file membership. Neither of these channels of communication, however, appeared to enable the union to transmit its messages effectively to the majority of its rank and file. On Saturday, 2nd May, the union organised meetings outside the gates of the St Helens factories and at four out of the six meetings the union did obtain majority votes in favour of a return to work. This, however, did not turn out to be the prelude to a resumption of work. On Saturday, 16th May, a ballot was held under the supervision of the local clergy in which a majority vote in favour of a resumption of work was recorded, but again the strike was not consequently concluded.

All these experiences confirmed the union's view that its policies possessed rank and file support. Despite this support the union was unable to organise a return to work because, it was felt, there were powerful opposing forces at work that were willing to use any methods to thwart the union. Firstly, the spokesmen felt that the union was faced with the problem that intimidating threats and open violence were being used by a section of the strikers in order to prolong the dispute. The union's local offices in St Helens were attacked on 29th April and had to be closed, resulting in the union losing its local communications base until it was able to establish temporary offices in a Labour club on 13th May. At mass meetings that they attempted to address union officials were confronted with what appeared to be organised resistance which was aimed at preventing them from gaining a fair hearing. After the factory gates meetings on 2nd May some members did return to work only to encounter ugly confrontations with groups of pickets. At one factory a group of employees was prevented from leaving the works for over an hour and succeeded in extricating themselves from the situation only by offering an undertaking that they would not attempt to return to work on the following day. The spokesmen felt that the union had encountered opposition that was willing to use unfair and violent tactics if necessary. Secondly, the union faced opposition from the RFSC. The union spokesmen did not believe that this committee was responsible for the incidents and threats of violence that had arisen during the strike, but considered the committee to be a separate force hindering the union's attempts to secure a return to work. The spokesmen believed that the resistance that the union was encountering from this committee was quite unnecessary and that the tactics that the committee was employing were patently unfair. They felt that the propaganda issued by the committee was extremely slanted and unjustly intended to discredit the union. The motives of the members of the committee were considered to be suspect. Members of the RFSC were felt to be interested mainly in reaping power and glory for themselves rather than furthering the interests of the Pilkington employees. Their main concern was considered to be either to take over the union organisation in St Helens or to smash the existing organisation and create another which they would control. It was felt that having experienced a taste of power the committee's members were unwilling to withdraw into the background and acknowledge the attractive nature of the terms that the union had succeeded in negotiating with the company. According to this view the committee was attempting to prolong the strike largely to satisfy the desires of its own members to exercise power. It was also felt that the RFSC was inspired by outside political forces and influences. Some of the union spokesmen believed that the committee had inspired the initial outbreak of the strike. They believed that the firm had been infiltrated by agitators who were inspired by political motives and whose objectives were to create economic disruption in Britain by depriving the country of its major source of glass. The union spokesmen who did not go so far as to allege that the committee had originally inspired the strike for political purposes did nevertheless express the view that political forces were supporting the committee with both advice and money. Amongst the union spokesmen it was firmly believed that there were 'reds under the bed'.

THE OPPOSING FORCES

These views that were subscribed to by individuals who were still playing active roles within the local branch about the nature of the opposing forces with which the union was faced, formed the basis of much of the union's own propaganda that was aimed at discrediting the role that the RFSC was playing in the strike and emphasising the extent to which its members had been subjected to unwarranted threats and violence. As we shall see, the branch failed to comprehend the real reasons why it was encountering such fierce opposition. Also the union held an exaggerated view of the extent to which its rank and file members sympathised with the union's policies. Consequently the union's propaganda only served to inflame still further the feelings of its opponents, and cut little ice with the rank and file since the statements that the union was making were clearly based upon a failure to assess the situation correctly. Yet again the union was adopting tactics that could not achieve their intended results.

Nevertheless, according to our spokesmen, as the strike progressed opinion within the union became increasingly convinced that it was being challenged by forces inspired by evil motives that were willing to resort to evil methods in order to obtain their ends. With these forces it was felt that there could be no compromise. Within the union it was genuinely felt that it was reflecting the wishes of its rank and file members even if some sections of the rank and file did not yet realise that this was the case. The policy of the union, therefore, could only be to hold fast to its position, to refuse to concede any ground, and to wait until events turned in a favourable direction.

An interesting feature of the accounts of the strike given by the union spokesmen was that at no time did they regard the firm as the main party with whom they were in conflict. Few criticisms of the firm were made by the union during the dispute. The strike had not been triggered-off by the firm's resistance to demands that the union was making. In the early stages of the strike the firm had been understandably hesitant before making a pay offer, but when the offer did come it was a reasonable one. From that point onwards the firm was regarded more as an ally than as an opponent in the union's endeavours to secure a return to work. To the extent that criticisms of the firm were made by the union spokesmen they concerned the firm's failure to be a sufficiently strong ally. For instance, one shop steward argued that the firm could have done more to publicise the favourable terms that it had conceded to the union. As a group, however, the union spokesmen did not regard themselves as being in conflict with the firm.

'Victory'

The manner in which the strike eventually ended was regarded by the union spokesmen as a victory for the union and a complete vindication of the firm line that it had taken against opposing forces during the latter stages of the dispute. As they saw it the strike collapsed as the rank and file gradually appreciated the reasonableness of the agreement that had been reached between the firm and the union, and as they gradually realised the extent to which they had been misled by the propaganda of the union's opponents. The majority in favour of a return to work that was revealed in the ballot supervised by the clergy confirmed their previous impressions that the majority of the members supported the policies that the union had been pursuing. From that point, according to the union spokesmen, despite intimidation and threats of violence, over 2,000 men returned to work on the following Monday and this figure gradually rose during the course of the week. At this stage the RFSC called the dispute off as a face-saving gesture; they knew that they were beaten: the offer of conciliation from the general secretary of the TUC was merely a let-out.

With the return to work its spokesmen believed the union had re-established itself in the eyes of its rank and file members. It was believed that the members now understood that the union had behaved properly during the strike, and that support for the RFSC had drained away as the

'VICTORY'

rank and file realised that this committee had misrepresented the union and unnecessarily prolonged the strike. As its spokesmen saw the situation the total pattern of events placed the union's role in the dispute in a favourable light. The union had not started the strike but once the dispute had begun within two weeks the union had used the situation to secure an increase of £3 per week on the pay of its members. The RFSC had succeeded in prolonging the dispute for a further five weeks, inflicting severe financial hardship upon the rank and file, and had gained nothing.

Once the men were back at work the union spokesmen believed that conditions would return to normal. Despite the strenuous attempts that had been made to discredit the union its spokesmen did not anticipate any widespread attempts to resign. They expected the members of the RFSC to attempt to take over positions of power and influence in the local branch, but the loyal shop stewards predicted that these attempts would result in abject failure. The members, the union spokesmen believed, now realised that the people behind the RFSC could not be trusted. Little further unrest in the firm was thought likely. It was felt that the rank and file had learnt a lesson. The men had experienced severe financial losses as a result of the strike and would accordingly think carefully before allowing themselves to be drawn into any further trouble.

Reforms in the union's branch structure and in the firm's negotiating procedures and wage system were expected to go ahead. The good that the strike had done was to speed up changes that would otherwise have taken longer to introduce. But in the union spokesmen's view these beneficial side-effects of the strike had been achieved during the first few days of the dispute. The prolongation of the strike had been unnecessarily damaging to all concerned.

To some people it may appear amazing that a large and ostensibly powerful organisation such as the GMWU should have so consistently misinterpreted the situation in which it was acting during the strike at Pilkingtons, and that it should have made such a series of tactical errors. We feel that certain key features in the union's situation can be identified which make this apparent ineptitude comprehensible.

Firstly, the union faced the problem that despite possessing an imposing and sound national superstructure, even before the strike its relationships with its rank and file members at Pilkingtons were weak. There was little rank and file involvement in branch affairs and communications between the branch and the grass roots were poor. The union's impressive superstructure was related to its sizeable membership in a decidedly tenuous manner, and in this respect the GMWU in St Helens was facing a problem that confronts large sections of the modern trade union movement. The tenuous links that bound the union's structure to its membership could easily be shattered in a crisis situation such as that which arose with the outbreak of an unexpected strike. It was easy for the union's superstructure to be left floundering upon shaky foundations, unable to assert any control over the situation.

A second point to bear in mind is that despite its apparently solid structure, decisions of crucial importance in the union were taken at branch level by small numbers of fallible individuals. Many of the individuals who played a part in determining the union's reactions during the strike had no previous experience of handling such a situation and had received little training that could have prepared them to deal with the circumstances that arose.

Thirdly, to union officials any wildcat strike immediately spells trouble. Their instant reaction to the situation is to feel that it must be brought under control. In the St Helens strike this was the initial reaction of both the local shop stewards and the union's full-time officers. As a large organisation a union inevitably develops a preference for ensuring that grievances are dealt with through formal negotiating channels. Furthermore, the long-term viability of its organisation and the security of its officials' positions depend upon a union persuading its members to channel their grievances through the established machinery. In addition, the involvement of union officials in negotiations with employers, and possibly also on government bodies, results in local grievances being seen in a different light to that which is evident to the members who are immediately involved. To the individuals who are caught up in wildcat strikes the issues involved will be of immediate and paramount importance, and the job of the union will be regarded as being to respond to the feelings of its members. Yet the reaction of the union official will be to define a wildcat walkout itself as troublesome and to feel that the situation ought to be brought under control.

Officials might attempt to control the outbreak of such trouble by adopting the aims of the strikers. But from the point of view of union officers such a course entails obvious long-term disadvantages. It would act as an incentive to its members to 'take matters into their own hands' on future occasions, and it would make established relationships of co-operation and trust with employers difficult to maintain. Hence in any wildcat strike it is always on the cards that a rift between a union and its members will become apparent.

This is a danger which is inherent in the structure of the contemporary trade union movement. In the light of this and the other considerations that have been outlined the problems that the GMWU created for itself as a result of its own tactics during the Pilkington strike must become more readily comprehensible.

Chapter Five

PILKINGTONS

The employer whose labour force is involved is normally regarded as a major participant in any industrial dispute. It is generally believed that the actions of an employer at all stages during a strike are bound to be critical, and that the policies that an employer pursues are bound to exercise a decisive impact upon the course of a dispute.

During the Pilkington strike, however, the role of the firm was hardly prominent; partly out of choice, but mainly because events forced the company on to the sidelines. The firm did little of decisive importance. For the first two weeks of the strike it was publicly refusing to negotiate prior to a return to work; an offer of a £3 increase on gross rates of pay was then made and accepted by the union. Thereafter the company appeared to be sitting back and allowing the strike to take its own course. The fact was that throughout most of the strike the company's hands were tied. There were few initiatives that were open to the firm. This chapter will therefore not be concerned with explaining what the company did, so much as why it was obliged to retreat into the background and, in the words of one of its senior executives, 'play it cool'.

When the strike broke out a Strike Executive was set up by the board consisting of five senior executives, which assumed responsibility for handling industrial relations during the course of the dispute. Regular strike bulletins were issued by this executive committee in conjunction with the group's public relations department. The bulletins were issued for circulation amongst the company's staff to keep everyone informed about the course of the strike as it appeared to the firm, and to explain the actions that the company was taking. In order to obtain a view of the strike from the perspective of the company we were able to gain access to these bulletins. We also had at our disposal the written evidence that the company submitted to the Court of Inquiry, and we were able to conduct tape-recorded interviews with two of the members of the Strike Executive after the dispute was over.

The Pilkington 'style'

In Chapter One we characterised the Pilkingtons as being austere men who had tended to endow the pursuit of their business with high moral purpose. We suggested that with regard to their employees they had developed a sense of responsibility for their welfare, while at the same time expecting from their workers the same devotion to the firm and to hard work that they themselves had. We showed, by examining various statements of prominent family members, that to some extent this nineteenth century zealousness continued to be of some importance for labour–management relationships in the firm in more recent decades. Whilst this tradition is by no means dead, the circumstances no longer exist which would enable it to thrive. There have been two changes which are weakening the tradition. Firstly, the vast increase in the scale of operations, particularly since the end of the Second World War, has meant that members of the family have just not been able to exercise the oversight over the running of the firm that once they could and have had to rely increasingly upon career managers who, no matter how much they might have imbibed the traditions of the company, will not have experienced the company as a *personal* thing in the way that family members have. This observation was to some extent implicit in what one of the senior executives said: ' . . . going back to ten years ago, a General Board Director would be seen much more in the works. Being the Works Manager's immediate boss, he would go down visiting every other day in St Helens; he'd probably say "let's have a walk round the works . . ."
I think top management have relied a lot on personal contact by living in St Helens, knowing the works and being round

the works, and this was a good way of keeping in touch. But because of the size that the company had grown into, we needed to replace this with amplifiers of what the shop floor was thinking . . .'

The second change refers to the workers themselves. This change too is partly a consequence of the increase in the scale of operations: it is difficult to think of oneself as a member of a family when one has 8,000 or so brothers and sisters and a few thousand more aunts and uncles. But it is also a consequence of the breakdown of the isolation of St Helens, and therefore a breakdown of whatever feelings there may have been of dependence on Pilkingtons for a livelihood. This was also reflected by one of the managers in reply to our question as to whether or not he thought St Helens was dominated by Pilkingtons: 'I think it is probably less true than it was. I think there was a time when we did, and I think that whether we do or not, people feel we do. There's a myth, if you like, about it.'

It is not however the habit of traditional ideas to disappear suddenly, and this applies to workers quite as much as managers – as we shall see in the chapter on the RFSC. Where Pilkington management is concerned strong vestiges remain. There remains an insistence, for example, that there is more to business than just making money:

'I think certainly the moral side does weigh with the company. If one runs a business one is to some extent one's brother's keeper. I think the company would still regard itself as being in business for something more than just making money, in the sense that it takes long-term views, and a long-term view is obviously that you have to look after your human capital as well as your money, and that it isn't just what you do this year that matters, but what you are working on is going to bear fruit in ten years' time. It is important that the company is not only profitable, but also has a "heart".

'During the '30's some (family) members put a lot of money into trust funds – so-called charitable funds – which are worth three or four million now in capital terms because they have grown. A lot of the welfare, i.e. widows' pensions, free coal and so on, comes from the charitable funds and not from the company as such. It has always felt right, and still feels right, that one should look after people as individuals and that one should try and provide for the total man if one can. A lot of it, I think, is historical, from the days when recreation clubs were "the thing" to have, and to have welfare workers was to be ahead of the field. These sorts of policies were very successful, and therefore one tends to carry them on, and I think there is a lot of it that is still basically right; that you should look after people, that you should try and encourage them to subscribe to a proper pension for instance, and that you should have people who can go out and visit them when they are sick.'

The strike did not, remarkably, shake this philosophy:

'I think we ... thought that we had, and I still believe we have, a tremendous fund of goodwill. The fact that we take the trouble, the fact that we think of people as individuals, the fact that we look after pensioners and this sort of thing, I think stood us in good stead and will stand us in good stead again.'

Yet on the other hand there are signs of change, for another senior executive told us that 'of course you can have something like teamwork':

'But the objectives of the various echelons in industry are genuinely different. Therefore, there is always going to be some time when conflict will arise. I really do believe that it is quite naive to believe that someone like a yard sweeper is willingly going to do something that is to his apparent or real disadvantage for the sake of the good of the company as defined by, perhaps, the Board of Directors.'

Furthermore: 'I think that (consultative relationships with shop stewards and trade union officials) can be pretty well organised, but it implies that the people who are consulting are reasonable people and will accept the inevitability of compromise because the interests do in fact conflict. Take the key problem – redundancy ... It is the objective of industry to employ as few people as possible: it is the objective of the unions to find as many jobs as possible. Now I know that in the long run these don't necessarily conflict, but there is no end to this long run.'

What this adds up to is that behind the united front that Pilkingtons presents to the world there exist contradictory philosophies as to the proper relationship between employer and employee. This is not to suggest that there are, within Pilkingtons, competing parties which enter into heated debate. The fact of the matter is more likely to be that if these differences do sometimes manifest themselves in arguments over policy, the various protagonists will not see that underlying the arguments are contradictory points of view. The strike however may well have made different people more aware of these conflicts.

The implication of the manner in which new attitudes have intruded upon the traditional Pilkington style is that one could not expect the company's management to have reacted in a totally united way to the outbreak of the strike. Some directors were shocked, whilst others took a more 'realistic' view. Likewise there were differences of opinion over what action the company should take during the course of the strike. As we shall see was also the case on the RFSC, there were 'hawks' and 'doves': there were those who wanted to take a hard line and those who wanted to be conciliatory. Because such differences did exist there must be an air of unreality about this chapter, since we shall necessarily be primarily concerned with reconstructing the public face that the company wore during the strike. The different points of view that this composite image conceals will have to remain largely off-stage. A further implication of the breakdown of the traditional Pilkington style that was already underway before the strike is that to the extent the strike has resulted in the firm re-assessing and changing its industrial relations policies, the groundwork for these changes had already been laid before the dispute occurred.

GUNPOWDER + ENGINEERING = STRIKE

Gunpowder + engineering = strike

Pilkingtons were not expecting a strike any more than anyone else, and they were taken aback by the way it snowballed right around St Helens: 'We were undoubtedly surprised at the time, certainly surprised by the rapidity with which it spread. The surprise was, to some extent, generated by the fact that whilst an excuse was made to have a showdown (this was the error in the bonus miscalculation) it very rapidly became a demand for half-a-crown an hour on the basic rate.' Management, again like most of the strikers, did not think the strike would last very long: 'I was certainly surprised that it had happened, that all those people that one knew had walked out. I thought "they'll go back by the end of the week". I was a bit hurt that this should happen, wondering why, and what one could have done to have stopped it.' 'Hurt' was almost certainly the predominant feeling amongst the Pilkington family as well for the firm was after all in ' ... the forefront of understanding the human side', and not just in business to make money. The firm's conviction was that ' ... people who had strikes didn't manage properly and did not understand about industrial relations and human relations ... There are some companies who are nowhere near understanding these things ...' This made Pilkingtons look outside the company for contributory factors. An explanation for the conversion of a dispute over bonus errors into a claim for a hefty increase on the basic wage was partly seen as the work of ill-disposed left-wing workers:

'I think to some extent the strike was engineered. I know the ground was probably fertile, partly because of the national picture, the prices and incomes policy having held back wages; ... I say engineered, though I don't imagine there was a central plot, but there are people working for us who are extreme left-wing people politically, who would naturally read the literature, go to conferences, and talk among themselves. I don't know what sort of network they have of linking with each other; Professor Wood, at the Court of Inquiry, asked "Can anyone explain how a bonus issue escalates to 2/6d an hour then to £10 a week, within two days?" '

In the early days of the strike, when hurt feelings and general puzzlement were at their height, conspiracy theories embracing revolutionaries out to subvert the national economy were more widely held than at the time we conducted our interviews. One rumour strongly circulating through Pilkingtons was to the effect that 'it is well known that two professional revolutionaries up from London are staying in the Fleece Hotel'.

There was also thought to have been some 'mass hysteria' present:

'We were surprised by the way it spread through the shifts including machine operators and tank people. I'm no expert psychologist and I don't quite know how these things develop. There must have been some mass hysteria about the thing. Certainly, there was evidence of that in the Sheet Works by the way the crowd suddenly descended upon the Rolled Plate Department, though the reception they got in Rolled Plate was less than enthusiastic. That was the department where the men insisted on shutting down their machines properly, taking their time and making sure that their equipment was in a safe condition before they left.'

Once over the initial shock, and under the stimulus of the Court of Inquiry, the firm began to engage in a strenuous process of introspection. Management began to wonder whether there had in fact been something wrong with the firm that had provoked the strike. The question was raised as to whether the firm could have taken forestalling measures to prevent the dispute from occurring. The company's negotiating procedures and its wages structure were both brought up for critical review. In retrospect the firm was prepared to concede that the conditions upon which it had been sitting amounted to 'gunpowder'.

'I would have said that a change had taken place over the last two years. Certainly our industrial relations side, in terms of odd disputes and walk-outs and this sort of thing, had apparently got worse. We had more troubles of that sort in the last two years than we had ever had; people told us, "you're sitting on a powder keg", but what do you do with a message like that, what does it mean?

'I think we had seen that we were at a point of having to take a step forward in our management–labour relations; we had started to see that the JIC was, as the central negotiating body, rather over-centralised with no sub-structure under it. We seemed to be getting less co-operation in change than we had had in the past. We knew the wage structure was getting out of gear.

'The other thing that we were aware of was that our lines of communication had increased tremendously. We divisionalised about three years ago, and I think that is one of the causes of the strike. It produced a lengthening of the lines of communication between the works managers and the board. Before the strike we had started to make progress on all these fronts. These lengthening lines meant that the works manager – in dealing with his people and the departmental managers – was further away from the inner councils of the firm, less involved in forming company policy, and less able to put it over. You see, going back to ten years ago, a General Board Director would be seen much more in the works . . . You have to communicate a feeling about working with each other; so to that extent I think perhaps people are further away and felt further away, particularly in St Helens.

'I know what did irritate the men a lot was that in a group of companies like this, there was an insistence on the part of the company that matters which had serious repercussions elsewhere had to be brought back centrally. That is time-consuming and we know that it was a source of irritation. It always was and always will be to the chap who has a complaint and who, through lack of understanding more often than not, thinks that it is local and therefore a simple one. It affects him and his mates and why the hell the manager can't say either "Yes" or "No" to it, he can't understand. And I don't think management explained this adequately enough although, God knows, we have tried often enough. I'm pretty certain the union didn't explain why things had to go back to the JIC so frequently. And, of course with the JIC things take time.'

In retrospect management was also prepared to concede that its wages structure prior to the strike had been unsatisfactory. Pilkingtons considered that the main problem with their wages structure had been not that average earnings were too low but that anomalies existed between the wage levels of piece-workers and those on the multi-factor bonus schemes. Consequently the firm thought that the £3 offer that was made during the course of the strike and which was applied to gross earnings, was much more equitable than the attachment of £3 to the base rate would have been: attachment to the base rate, on which piece-workers' pay was based, would have widened the gap between the workers on the two systems rather than have narrowed it. Since the firm had a longer term in view in which they were hoping to recast completely the whole wages structure the award was, from their point of view, eminently rational as well as generous: they produced figures to show that hour for hour there were only two other British industries who had better paid workers.

'We had perhaps not been insistent enough at earlier negotiations in the company's attempt to redress the balance between the two sorts of bonus workers we have. There was undoubtedly a split in the labour force. The direct system bonus workers were considered by people on the multi-factor bonus schemes, to be having it relatively too easy. This is nothing new ... Certainly, the highest earning people here are direct system bonus workers (i.e. those on piece-work), but their rate of earning is very much in their own hands and some of them are prepared to work extremely hard. In addition to which, bonus values have been in the course of 30-odd years steadily eroded, and of course no union is prepared to negotiate downwards to what would have been revealed by restudying. But the spread of earnings among direct system workers is very wide and the idea that a man on system automatically earns more than a man on the multi-factor bonus is not by any means generally true.'

Although it conceded that its wage and negotiating structures did create problems, management was quick to point out that most large firms, including nationalised industries, faced similar difficulties. The firm could not accept that such malfunctioning as existed justified the vilification of the company and the labelling of it as the guilty party in the strike. The negotiating machinery and the wages system were already under active review before the strike. Whilst, in retrospect, management was willing to admit that there had been faults in the firm's organisation, the feeling was never totally dispelled that the company was really an aggrieved party. Hadn't they always paid a great deal of attention to 'the human side', hadn't they always tried to be fair and considerate towards their employees, hadn't they a long history of catering for 'the whole man', hadn't they established feelings of mutual regard between themselves and the trade unions? It was felt that a certain amount of engineering by left-wing influences sufficient to provoke a state of mass hysteria was necessary in order to set a spark to whatever gunpowder may have been smouldering. Certainly, at the time when the strike broke out, the firm was taken completely by surprise.

Playing it 'cool'

Once the strike was underway, and throughout the course of the entire dispute, the firm had little alternative but to 'play it cool'. Actions initiated by other parties repeatedly tied the firm's hands, giving it little option but to remain in the shadows. For the first few days of the strike Pilkingtons were quite as bemused as everyone else. The company statement of 6th April said: 'The company has still not been told what the strikers want – unofficial reports from the various mass meetings are widely inconsistent.

The employee's side of the JIC is meeting tomorrow (Tuesday) and after that meeting, it may be possible to see more clearly what the purpose is' (*Pilkington News Bulletin*, 6th April, 1970).

As they were not sure what the strike was about, Pilkingtons could be nothing other than perplexed. What action could they take if they did not know what the strikers' grievances and objectives were?

The tactics of the GMWU did not make it any easier for the firm to define the problem with which it was confronted during the early stages of the strike. Whilst the local branch was declaring the strike official, the union's national officer was assuring the firm that it was totally unofficial. Yet whilst insisting that the strike was unofficial, the union's national officer was also suggesting that a pay offer would be necessary before a return to work could be organised. What was the company to make out of all this? The union that it recognised as representing the strikers was not squaring up for a showdown, hence there was no officially recognised enemy for the firm to attack publicly. The firm could not repudiate any claim since none had officially been placed before it. At the same time, the company could not put on too great a display of intransigence since union officials were hinting that a pay award would prove necessary. So what did the company do? 'We adopted the classical initial stance, of course: "You've got to go back to work before we negotiate." That's a different position from: "You've got to go back to work before we talk." And we talked to officials of the GMWU who had not recognised the status of the strike as official. We were talking to them from the start. "All right, we will meet you, using the normal negotiating procedure." '

During the opening stages of the strike, the firm could do little else apart from talking with union officials behind the scenes. Since the union's national officials were declaring the strike unofficial, the company could do little other in public except adopt the 'we will not negotiate . . .' stance. To some extent, Pilkingtons were reinforced in their decision to take

PLAYING IT 'COOL'

this initial stance by the fact that when they asked other firms what they had done in similar circumstances, Pilkingtons were told that a refusal to negotiate prior to a return to work was the normal line to take. Pilkingtons later discovered that their advisors had not, in fact, always acted in such a high-principled manner. But during the first two weeks of the strike no other policy was really possible.

The behaviour of the union during the early stages of the strike left the firm completely perplexed. 'The shop stewards were out of their depth, I think, a bit like we were. The stewards were not used to this sort of thing; they didn't know what sort of tiger it was they were riding. They tried to regain control on the sensible basis that if you can't beat 'em, join 'em. There was a great deal of confusion in the early stages where the seeds of this rank and file committee were sown. Some of the union officials here, as I understand it, made what in retrospect looked like a take-over bid for the dissident committee which was forming, and made some slightly wild statements about it being "official at local level" which is a meaningless phrase apparently according to the union rule book. I don't know whether that move by them – which failed – was good or bad, damaging or helpful. I imagine that in the long run, it was not helpful in that it put question marks against the credibility of some union officials.'

Whilst initially the behaviour of the union appeared schizophrenic to the firm, before the strike was very old Pilkingtons realised that the GMWU was facing serious problems of its own.

'It didn't take very long to recognise that although we were negotiating through the normal ... channels, there was very serious doubt as to whether in fact the GMWU represented the shop-floor. It was very quickly recognised that they themselves were under fire. How can you negotiate with a body of people when they are themselves involved in a pretty vicious civil war? It isn't obvious and certainly isn't easy. I don't know that you can do anything other than play it cool. A lot of the ammunition being fired by the RFSC at official union representatives took the form of accusations that the GMWU was the "bosses' union". It is not up to the company, it is not common sense for the company, in that sort of situation, to take a position which gives the rank and file any ammunition to say, "There you are – we told you it was the bosses' union." You have got to play it cool. I don't see what other action we could have taken because, quite honestly, some mornings we wondered whether Pilkington Brothers had anything to do with this strike or not. The events of the day frequently appeared to have had nothing to do with us, but to be a battle about who was going to be transferred from which side to which.

'I think it was very difficult to know what to say. In any case, anything you might say may make the situation worse ... There was the feeling that it was the right thing not to say too much particularly in the early stages; one thought that people would come back, and one didn't want to say too much that might excite people or produce the wrong reaction ...'

The civil war that began to break out early in the strike made it increasingly difficult for the firm to play a positive role in the dispute. Whatever approach they adopted towards the union, conciliatory or aggressive, appeared likely to have the effect of inflaming the situation. But whilst the firm found the union's early handling of the strike rather perplexing, management later had to admit that:

'They (the GMWU national officials) gave us the right signals from the start: they told us we would have to negotiate.'

And negotiate was what the firm eventually did. Whilst on 13th April at an emergency meeting of the JIC the firm repeated its formula of no formal negotiations without a return to work, by 17th April the firm had changed its mind and was offering to improve, as an interim measure, an increase that was due to be paid on 2nd May of $3\frac{1}{2}$d an hour. The union declined this offer, and three days later managed to persuade the company that an increase of £3 would probably suffice to get the men to return.

PLAYING IT 'COOL'

The firm had changed its mind because:

'... it didn't take very long before it became obvious that there wasn't going to be a return to work to enable negotiations to start and therefore we had to face a factual situation. In those circumstances you have to come round to negotiating whether you are under duress, in the classical sense, or not.'

Having publicly announced a pay offer the firm might have brought itself to the forefront of the action. However, events were once again to push the company on to the sidelines. Once its offer had been accepted by the union, but immediately rejected by the rank and file strikers in St Helens the firm was placed in a quandary. It could not improve its offer since the union that the company recognised for negotiating purposes was not asking for any more. The firm could hardly criticise the union for failing to demand more generous terms. Nor did the firm feel that it would be wise for it to publicise the generous features of the offer that the union had negotiated: such a course could have played into the hands of those who were claiming that the GMWU was hand in glove with the bosses. At the same time Pilkingtons did not want to launch a vociferous attack on the unofficial strike leaders: the union's position looked insecure and the possibility had to be entertained that eventually the firm would have to negotiate with the people who were then outside the factory gates. The firm did not want to inflame still further the feelings of those who were critical of the recognised union. Furthermore, Pilkingtons believed that if they bewailed the losses that the company was suffering due to the strike this news would bolster the militants' spirits. Any indication that the firm was weakening or that the strike was biting to the bone had to be avoided. Pilkingtons even requested some of their major customers not to make too much noise about the strike's dire consequences for *them*. In view of all these considerations the firm felt that it had little option but to sit back and wait for events to take a favourable turning. The firm seemed to be quite lost:

'I think it was Mr X who said, "I seem to fluctuate between suicide and euphoria," and that was about the situation. "What is the situation this morning?", "How many shop stewards have they got?", "What is the state of the battle?". There were umpteen mass meetings, forecasts of "Oh yes, they'll vote today – they'll come back". They didn't.'

Throughout the strike the firm continued to talk with and maintained friendly relationships with the union. Both parties had the same objective: to secure a return to work. If relationships between firm and union did not amount to a conspiracy the former saw no reason why it should not perform certain services for the latter. Yet whilst remaining in contact with the union the firm had to consider the possibility that it might eventually have to talk with the RFSC.

'We considered it – we considered every conceivable possibility. But there was never any clear evidence as to who represented our employees. We have got bits of paper which say it was the GMWU. The evidence fluctuated. It is my impression that the number of loyal shop stewards, if you can take that as some measure of support the GMWU had, dropped at its low point, into single figures. At this stage we were wondering how much longer they would survive. But they were there and they did come and talk to us. Had they not survived, then I suppose there would have been no question: we would have had to talk to somebody, who I presume could have been the RFSC.'

The nearest the company came to attacking the RFSC was in its letter to employees on 12th May.

'The unofficial strike committee has one thing to offer you – that many of you will have no jobs to go back to.'

This was the strongest public assault that the firm launched against the RFSC.

Pilkingtons clearly wanted to keep all the options open for they were aware of the possibility of having to talk with the RFSC: it would not have been wise to launch a frontal assault on people with whom it might have to talk. Pugnacious statements (that some members of management would certainly have liked), apart from jeopardising future bargaining relationships, may also have hardened the attitudes of some rank and file strikers. This had to be avoided. In addition Pilkingtons did not want to do anything that could create a reservoir of ill-feeling towards the company and thus inhibit the establishment of faithful co-operation once a return to work had been organised.

The 'unofficial people'

If the firm's public comments about the RFSC were muted this does not mean that management did not harbour private feelings towards them. Naturally enough, the firm's attitude towards the RFSC were not generally friendly, but management was prepared to concede a point or two.

'I think they were pretty sincere in their dissatisfaction with the negotiating procedure.'

And there was at least an implied sympathy when it was said of the GMWU:

'It is easy enough to criticise other people, and it is not my job to criticise the union, but I feel they were a bit insensitive. I think the reason they were that way was that they had not taken the same attitude as the company. The company as it grew made a criticism of itself which was that we were in danger of creating too many administration people. If this union branch were to be criticised it would be that they had far too few for the growth of numbers.'

In terms of the strategy of the strike the RFSC was seen as a power grouping with a strong basis of support and that was sufficiently well-organised to put on a formidable show of picketing strength to keep out those who wanted to go back to work:

'They were making a power bid to take over the negotiating position. I suppose it is not surprising. A lot of their tactics, at the time, looked clumsy (a polite word), and the wrong tactics to secure their objectives.'

And: 'I think the mechanics were fairly clear. You had a

group of people who were sufficiently powerful to persuade quite a lot of people that if they stayed out they'd get more, and were sufficiently powerful to "persuade" those people who wanted to accept the agreement not to go back to work.'

Pilkingtons were not convinced that the RFSC's disavowal of 'violence' rang with truth and sincerity; they were more than a little suspicious of their intentions. As one of the directors said in the closing stages of the Court of Inquiry:

'If the unofficial strike committee are sincere in their declaration that they deplore violence, then there is a simple solution to this matter of deciding who really wants a return to work. I give the unofficial strike committee a challenge, and this is to take off all the pickets so that everyone can make up their minds without any pressure.'

If the firm was prepared to concede a further point – that the RFSC had some members of ability – it was also prepared to think that at least some of its members were applying that ability for political ends:

'I would have thought that some of them would be extreme left-wing, that is people who, on the whole, did not believe in the capitalist system, did not believe in the present society, and therefore were working from a very different base. Certainly one had the feeling that they were very able – some of them anyway – as the thing went on, to be able to organise it as they did.'

Once the strike was over the firm's suspicions of the RFSC did not have to remain muted and an early opportunity was taken to crush the breakaway union.

'Good and reliable men'

One of the favourite parlour games in Pilkingtons was guessing the number of their workers who were in favour of the strike in the first place, and in favour of continuing it after the £3 offer. The initial feeling had been one of disbelief that a majority could have been in favour; were they not mostly good and reliable men?

'GOOD AND RELIABLE MEN'

'I would think about (our employees) as being mostly good and reliable men with wives and families, who come in faithfully to do their work day after day. Of course, I knew a lot of them; I worked with them in the early stages when I came into the industry; they were basically hard working, responsible people. But one also knew that the turnover was increasing in certain areas: one had a lot of people who just came and went and weren't really interested in the industry or anything except picking up the money.

'I didn't think the majority would support it. I thought they would go back. I was very surprised when they didn't go back after the £3. I certainly think after the £3 that the majority did not support the strike. I was surprised at the result of the ballot at the end of the sixth week, I thought it would have come out more in favour of returning to work.'

Another senior executive was more analytic:

'What intrigued me was that the labour force appeared to break up into three almost equal parts . . . We never had more than an equal number, I think, who were obviously wanting to come back to work and willing to have a go at the picket lines. The other 3,000 disappeared into the blue. What they wanted to do can only be deduced as not wanting to get involved. I certainly believe they weren't prepared actively to dissent from the official union position. Equally, they didn't seem to be prepared to take any risks of walking through the picket lines to come back to work. I believe they were either apathetic or apprehensive of becoming active on either side. So whether the majority wanted to come back to work, I don't think you can say. Among the active people, they were about equally split . . .'

As the strike drew on the firm's faith that *its* workers could not *really* be supporting the strike gradually dissolved. The management of any firm that is involved in a dispute must spend part of its time trying to guess the volume of support for a strike, for a reliable estimate will indicate the price that will have to be paid for a return to work. Yet in the case of Pilkingtons the guessing game was of more than tactical importance. To many senior managers a troublesome question was whether all the firm's 'progressive' labour relations policies, and all the goodwill that they believed had been built up in St Helens really counted for nothing.

Climax, anti-climax, climax

Pilkingtons played little part in bringing the strike to an end. As with so many other events during the strike the firm watched the concluding scenes being acted out as an interested spectator. The firm had no prior knowledge of the GMWU's intention to organise the 'parsons' poll', but were favourably impressed as the union's tactics unfolded.

'The union were coming up to this ballot of course, which they did not tell us about; we only saw it on television like the rank and file people: that was the first moment we heard it. It was a very astute move.'

After the result of the ballot had been declared it was Pilkingtons turn to feel elated. They had already prepared a letter to send out to their employees informing them as to how they should go about reporting for work. Many office workers were busy that weekend delivering the letters. But the moment of triumph went a trifle sour on Monday, 18th May, when only slightly more than a quarter of their workers turned up. By then of course arrangements had been put in hand to ensure that a large number of police were available to protect people who went back. Pilkingtons had accounted for the possibility that the poll would not assure a complete end to the stoppage for the prepared letter stated, 'The assurance ... about maximum police protection on entering and leaving the works has been reconfirmed.' When barely over a quarter of their labour force turned in for work on the following Monday the 'parsons' poll' turned out to have been an anti-climax from the firm's point of view.

There was, however, a further strategic move to end the strike for the firm to watch. This was the intervention of the General Secretary of the TUC. Again, Pilkingtons were impressed by the sense of tactics that was being displayed.

'I thought that was the right answer. Everyone is looking for face-savers at some point, and I thought that Vic Feather gave the required face-saver for the unofficial people to take. I was surprised when they didn't take it, but Vic Feather was smart enough to follow up with a second telegram which they did take.'

This time the trick worked: the strike was over. The manner in which the dispute had developed into a civil war amongst the labour force had reinforced the firm's view of itself as an aggrieved party in the strike. Events forced the firm into a position in which it was being seriously damaged, but from which it appeared unable to organise its own means of escape.

Pilkingtons, like the GMWU and the RFSC, were aware that the end of the strike had not signalled the end of the battle. Indeed when we first approached management we were asked if we were not being a little premature. We were told: 'The men may be back to work, but the consequences have yet to work themselves out.' For the firm the end of the strike really signified the beginning of its involvement in the battle. During the aftermath of the dispute the firm had to start to repair the damage, both to its trading position and its system of industrial relations, that had been created by the strike. It also had to decide what to do about the breakaway union.

Chapter Six

THE RANK AND FILE STRIKE COMMITTEE (I)

The emergence of the Rank and File Strike Committee

Neither the RFSC nor any part of it existed before the strike. If there had been fairly widespread discontent with the union, it had never crystallised into any sort of popular movement with an identifiable group of leaders: it took a strike to accomplish this. A strike, like any other crisis, throws a number of pre-existing features of a situation into a sharper relief. People start talking and thinking about things that they have never really thought or talked about before. Or if they have thought and talked about them it has been in a desultory, abstract way – the way people talk when there is no sense of urgency, no possibility of action.

In the everyday life of factory work the union is not an organisation that could be said to have any great grip on the feelings and imaginations of people. The union is an 'insurance policy' on which one falls back in time of need. If it amounts to a popular movement for the politically conscious shop steward and full-time official, the same could hardly be said for the great mass of the union's membership. Other things apart, the organisation of the unions is not conducive to perceptions of them as mighty warriors in the 'workers' struggle'.

The very great majority of trade union members are clients – they appear to behave as clients and are treated as if they are clients. But this is to oversimplify somewhat for there do remain, amongst rank and file trade unionists, residues of earlier attitudes. Attitudes, not quite buried, which suggest that the trade unions are *our* organisations, that they *ought* to be responsive to our wishes. Thus, even if the trade unions have become *bureaucracies*, there remains a feeling that they ought to be *democracies*.

THE EMERGENCE OF THE RFSC

A strike inevitably brings these sorts of feelings bubbling to the surface. Other bonds have been temporarily cast aside. The union comes to assume a prominence in people's minds that it rarely enjoys: it becomes of overriding importance because it is the only visible means of support.

Now this changed response to trade unionism creates very real problems for the full-time officials. They have grown accustomed to the client relationship between themselves and their members, and to unchallenged authority and are therefore rather perplexed when suddenly people are unprepared to accept their advice and do as they are told. This attitude of officials is further compounded by the fact that the official's way of life is not his member's way of life. The chances are that he spends much more of his time in the company of employers, government officials of one sort or another, and with local dignitaries. The natural result of this is that he tends to develop interests which may sometimes go against the interests of his members, and he will certainly find himself saying to his members: 'If you knew what I know you would appreciate the difficulties involved in what you propose.' Very often of course he will be right, but even when he is the man on the receiving end of the pronouncement he will be thinking to himself: 'What you are really saying is that if you go out for what I want you're going to create problems for yourself in your relationships with employers and government officials, etc.'

In an unofficial strike then there is always a real possibility of some sort of revolt within the union, always the chance that an unofficial body will assume control and turn on the union. That this happened at Pilkingtons should not have surprised any experienced trade union official.

*

Discontent and distrust had been expressed by the rank and file strikers right from the beginning, although it appeared to have been directed as much at shop stewards as at full-time union officials. It was not exactly diminished by a national officer of the GMWU when, at a mass meeting on

THE RANK AND FILE STRIKE COMMITTEE (1)

the sixth day, he told the strikers that they did not have union backing and that they should return to work. This message was not well-received – he was booed, jeered, and shouted down. Neither at this nor at any subsequent stage did the GMWU give any indication that it was prepared to attempt to use the strike to get an increase of pay out of Pilkingtons. Little wonder then that the vast majority of the rank and file, knowing little of the politics of trade unionism, should have developed strong anti-union feelings.

The time was ripe for the formation of an RFSC. It is easy to state the obvious and say that the RFSC was a response to a popular, if unorganised, protest at what seemed to be a wilful disregard of the members feelings on the part of the union. But such a statement says nothing about how the RFSC actually came into being.

The reconstruction of the 'birth' of the RFSC is not a simple matter because the people involved were very vague about it. As one of them said: 'It just sort of came from nowhere.' And this is really the truth of the matter for there was no question of a number of discontented people meeting surreptitiously in a pub, and deciding that a RFSC was needed. But if it emerged out of 'nowhere' in the sense that it was not organised, it did not emerge out of 'nowhere' in the sense that it was not a culmination of certain peoples' actions. It did *of course* result from peoples' actions, it was just that nobody saw at the time what the consequences of those actions would be.

On the first Monday of the strike (6th April) two people who were eventually to become prominent members of the RFSC secured the loan of loudspeaker equipment for the union. On the next day the Triplex shop stewards set up their own strike committee as did all the other plants. It was out of these two unconnected and innocent events that the RFSC was eventually to be born.

The point of mentioning them is this. Charlie F and Dave B, the two men who were so active on the union's behalf, were vigorous and able men who were deeply suspicious of the union. Neither of them had particularly wanted the strike, but once it was on they were determined to do all they could to make sure that the union exploited the position of strength and put the 'screws' on Pilkingtons. They did not however have an organisational base: they were, as far as they knew, two isolated individuals. They knew that there was discontent both amongst the rank and file and amongst branch activists, but they didn't see any signs of a coherent opposition developing.

The Triplex committee was soon to provide the organised base as well as other men of ability – their chairman was, as the GMWU branch secretary was later to say, '... one of the finest union representatives on the shop floor in this town'. The Triplex stewards had been up against the union for some time – they had resisted being brought into the Joint Industrial Council when Pilkingtons had assumed control of Triplex some years before. They also had an agreement with Triplex management that no full-time union officials were to be allowed into the plant without the agreement of the works convenor. Furthermore, some two months prior to the strike they had proposed to a special branch meeting of the union that Triplex should break away from the JIC and seek parity of base payment with the Birmingham Triplex factory organised by the Transport and General Workers Union. The proposals were voted down, whereupon the Triplex stewards decided to go it alone some time after 2nd May when their agreements expired. Triplex, in short, was already geared up for action.

Thus on the one hand we have two vigorous men, suspicious of the union but isolated, and on the other an organised group, competently led, which had already been in dispute with the GMWU. Twelve days were to elapse before these two groups came together. In the meantime unrest continued to grow amongst these activists: they complained about the lack of organisation; they complained that the union's central strike committee had been picked rather than elected; about the union's refusal to make the strike official; about its apparent lack of aggressiveness towards Pilkingtons; about its failure to keep the rank and file

THE RANK AND FILE STRIKE COMMITTEE (1)

informed; about the general air of despondency amongst some of the older GMWU men. These complaints were often voiced at the daily meetings in the GMWU offices.

The GMWU meetings provided the opportunity for the activists to meet and commiserate with each other. By Thursday, 16th April, Charlie F and Dave B had despaired of the union and decided to call a mass meeting for the next day. They told the Triplex men what they had decided and asked if they were interested in joining them – they were. That night they went around town sticking up posters and handing out leaflets advertising the meeting.

At the meeting on Friday, 17th April, the crowd was told that the platform had no confidence in a GMWU national officer: 'At all our meetings he has never mentioned money – only a return to work. We want you to give us the power to go in and make wage demands ourselves' (quoted in *The Lancs. Post & Chronicle*, 17th April). A platform speaker asked the meeting for representatives to come forward from each factory so that a committee could be formed, and someone from the crowd shouted: 'You form a committee Tom, we know the people you'll choose will be all right.' After the meeting, which had voted to continue the strike, about 1,000 strikers marched to the Pilkington head office where a meeting between the union and Pilkingtons was taking place. A delegation from the newly formed committee attempted to see Lord Pilkington: they were rebuffed.

The mass meeting had been well-timed in two respects. In the first place the organisers knew that the union was meeting Pilkingtons that day which meant that a good show of force by the rank and file might put some 'ginger' into the union. And in the second place it gave the organisers a chance to urge everyone to vote to stay out on strike in the postal ballot which had just been announced by the union.

After the demonstration at Pilkingtons broke up the committee went to one of the local Labour clubs, elected its officers, and decided the composition of its negotiating committee. The RFSC was born.

WHAT SORT OF MEN?

At this stage there is little doubt that the RFSC saw itself more as a 'ginger group' than anything else for it was not until three weeks later that it organised mass resignations from the union. At this time all that they wanted to replace was the GMWU negotiating body, and not the union itself. Other RFSC activities in the ensuing week support this contention – such as the march on the GMWU offices on 29th April to renew demands that the strike be made official. Not until the appearance of the GMWU coffin in the May Day parade did it become evident that a civil war was under way.

What sort of men?

Most groups of men can be relied upon to have in their number a wit, a musical entertainer, a dreamer and romantic, a tough realist, and a shrewd tactician. The strike committee had all these types and more. Most of them had their secret 'ambitions' too, although none could have been called embittered or frustrated. Even the most ambitious members were to some extent scarred with that fatalism so characteristic of working class people – we were frequently reminded of that Brazilian peasant saying: 'If the poor could shit gold, they'd be born without arse-holes.'

*

George F[1] banjoed vintage Lonnie Donnegan down the escalators and along the platforms of Tottenham Court Road underground station followed by a singing strike committee. The passengers on the train joined in and nobody gave up until the top of the escalators at Euston station. That was after 7½ hours abortive negotiating at the TUC.

Charlie W, a reader of Shelley and a quoter of Omar Khayam: 'The ball no question makes of eyes or nose, but right to left the player goes, and you cast it down into the field . . .': 'his twisted sense of logic appeals to me'.

---

1. These names are fictitious.

THE RANK AND FILE STRIKE COMMITTEE (I)

George F, who has taken evening classes in English and who wrote essays on pollution long before it became a fashionable topic: 'I wanted to be a writer about nature like Henry Williamson – I think he's great.'

Jim J, a vigorously loquacious man: 'I was never clever enough, but I'd loved to have been a lawyer. Fighting for people's rights and things like that – I'd have loved it.'

Arthur H, a quiet-spoken and thoughtful man: 'I would have liked to have been a person like Jack Dash, a person I really admire. He pushes convention to one side and goes for something heart and soul. He has the people behind him and he doesn't let them down.'

Harry J, quiet, neat, and methodical: 'I would like to have been a construction engineer doing road and bridge building. But this was chopped years ago when I had to leave school. You had to have money to go to the grammar school and onwards in those days, but my dad was out of work for five years. I was brought up on jam butties.'

Nick M, a restless young romantic: 'I'd really like to travel and throw off the bonds of convention. To ramble, to go hither and thither like the wind. I'd like to see how I would stand up to the emotional barrage of being out in the world alone.'

Graham D, an ex-white collar worker full of nervous energy: 'I'd like to have gone to grammar school and to university and be more knowledgeable – I'm tied to certain subjects. I can't talk about politics and world affairs which I'd like to do. But obviously I wasn't clever enough so that's it.'

Simon P, an intense man who carefully assembles his thoughts: 'I'd like to have been a doctor. I have a feeling for people that makes you want to do something for them. I think as a doctor you accomplish something like an artist does.'

Phil T, a family man: 'I achieved my ambitions with my wife and three kids.'

Roy F, blunt, plainspoken, as honest as the day is long: 'I'm sorry I never had an education, but I don't think it

WHAT SORT OF MEN?

would have changed me as far as looking after my fellow man. I don't think anything would ever change me, educationwise or any other way, of going against the working man. I think I'd have used it for his benefit.'

Sam T, a young rebel if ever there was one: 'I just take life as it comes. If something comes my way and I think it's worth fighting for I'll fight for it.'

Arthur G, withdrawn, but political to his toe-nails: 'I would have liked to have been in a position where I could initiate large changes in our society, particularly in education. I would like to be the architect of changes whereby people could control their own lives.'

Tom P, a listener, shrewd and wily: 'I've always felt it in me that I could write a book. I tried it once about ten years ago. I really felt I had something to tell somebody. I still feel that. I'd write about my experiences because I think I've had a bloody good education.'

All of the RFSC were Lancashiremen. All except three had been born in St Helens and the others had been born within a twelve-mile radius of St Helens. They were neither notably old nor notably young: if most of them were in their mid-thirties and forties, there were three in their fifties and four in their early twenties. The youngest was twenty-two and the oldest fifty-six. Every one of them had left school at the minimum age although two were ex-grammar school boys and one an ex-public school boy.

There were a couple who had been rebellious in their youth: conscripted during the war to work in the pits as 'Bevin-boys', one refused and was jailed for twenty-one days while the other absconded from the Durham pit to which he had been sent. Nearly all the others had been unsettled youths – four had served apprenticeships in now obsolete trades, but most of the remainder had frequently changed jobs within the first few years of leaving school. Three members at least had been offered promotion by Pilkingtons at various times and one had been a staff foreman.

Two-thirds of the RFSC had worked at Pilkingtons for

THE RANK AND FILE STRIKE COMMITTEE (1)

five years or longer – nine of them for between fifteen and thirty years, six for between five and fifteen years, and nine for less than five years. Of the short service men two were ex-miners, one had worked in the pit for thirty-four years and the other for eighteen: both had held office in lodges of the miners union. Slightly more than half had had relatives working at Pilkingtons at one time or another. One had his father, mother, and three brothers at Pilkingtons; one's father had worked there for fifty years, while another's had been with the firm for thirty-four years, and his grandfather for fifty-four years.

Just over half of the committee either were, or had been, shop stewards at Pilkingtons: one was works convenor at the Triplex plant, and another had been the senior shop steward at the Ravenhead plant. Nine members of the committee had relatives who had been active trade unionists. One of them had three brothers who were shop stewards, one had an uncle who had been a full-time official in the miners union, while another's father had been a noted miners militant in the earlier years of this century. Two members of the committee were elected as Labour councillors during the course of the strike – both of them having served previous terms of office. The RFSC was Labour to a man although only five were party members. They were almost Catholic to a man too, but nearly all lapsed Catholics. One of the few practising Catholics had a son training for the priesthood.

All but a handful of the RFSC had never been on strike before, and of those that had none had played a prominent role. Their inexperience was there for all to see – when one journalist remarked to us: 'They've got a long way to go,' he summed them up very well. But if collectively they lacked experience, as individuals each had some particular ability or talent to offer. Many of them surprised themselves. It was not just that they found themselves doing things they had never even dreamed they would do, rather that some of them were discovering abilities in themselves that they didn't know they had.

The strike for most of the RFSC was then a liberating experience, because for the first time in their lives they had cast aside all the bonds that constrain most people in their everyday lives. They were their own masters, for instead of having their actions largely dictated by a system of authority in which they were subordinates, they were now in a position to decide for *themselves* what they were going to do. From being in a world where their actions were mainly dictated by other people, they were now in a world which was much more evidently of their own making. This, essentially, was what George F was saying when he commented: 'I enjoyed the freedom it gave me to express myself which is something entirely new to me after working in factories.' But it was a freedom to *do* something: 'People have become free to fight a system that they know is wrong. People have got it off their chests, they've felt free,' said a man who has worked at Pilkingtons for thirty years.

Strike organisation

For the first two weeks such organisation as existed was more or less in the hands of the union. From the first Monday of the strike a central committee had been established and it met daily in the union offices. Strike committees had also been set up on a plant basis with varying efficiency – only the Triplex committee seems to have been organised on a formal basis: it had daily meetings in one of the town's Labour clubs. The Sheet Works committee met in the bicycle sheds at Sheet Works! The other committees seem to have been *ad hoc* affairs, meeting in a corner of the union offices. Picketing was minimal and ill-co-ordinated, and there were no continuous attempts to keep the rank and file informed of events. It was not until the RFSC took control that any coherent form of organisation appeared, and even then it was not until the fifth week that it started to look really efficient.

*

THE RANK AND FILE STRIKE COMMITTEE (1)

Before a strike committee can become recognisably powerful it has to attain one main objective – the solidarity of its supporters. And to achieve this a number of activities must be more or less efficiently carried out. In broad terms solidarity is achieved by effective internal organisation and by demonstrating that the propaganda battle waged on behalf of the strikers is achieving results. Above all else splits within the ranks have to be avoided, for if they are not they are open to exploitation by the 'enemy'.

The maintenance of solidarity was what preoccupied the RFSC: any sign of a break or a potential break in the ranks was immediate occasion for anxiety. It would be fair to state that the committee was in a perpetual state of anxiety, and this accounted for the fact that towards the end several prominent members of the RFSC were in an almost total state of physical and mental exhaustion.

Until the fifth week organisation was rudimentary and largely confined to organising mass meetings and issuing press statements. At a tactical level there had been attempts, through the Mayor, to get negotiations going with Pilkingtons, but the bread and butter of strike organisation – picketing, leafleting, 'blacking', organisation of financial support, etc. – was not thoroughly organised until the joint Pilkington–GMWU move to get a return to work in the fifth week.

Once the RFSC had established a permanent HQ in the Cotham Arms in the town hall square (rent paid by building workers on a nearby site) the division of labour within the committee took on a distinct shape.

The overall size varied from twenty-five to thirty people. One plant, Triplex, tended to be over-represented – it contributed one third of the membership, the other plants were represented in more or less equal proportions. Twelve of the general committee formed an 'inner cabinet' called the negotiating committee: it consisted of the main officers of the RFSC plus seven others. Most – but not all – major issues were discussed by them before they were discussed by the committee as a whole. On those occasions when issues arose needing immediate decisions they were made by those

STRIKE ORGANISATION

members of the negotiating committee readily available.

The RFSC met every morning at 9.30 for about 1½ hours. Reports were made of the activities of the previous day and decisions taken concerning future short-term activity. There was never any full and open discussion of tactics, though it clearly lay just below the surface of any proposition that was made, and was implicit in actual and proposed activity. The last item of each morning's business was to prepare a press statement.

For the remainder of the day the committee room was manned by a sub-committee of about six members called the inlets and outlets committee. Their job was to deal with a wide range of queries from rank and file strikers – anything from Social Security payments to stalling HP creditors. The amount of advice given was considerable and usually, in our judgement, sound. They received deputations of shop stewards from other firms who came with expressions of support backed up with cash and a pint of mild. They were responsible for organising 'blackings', and for addressing meetings of workers and students in the Liverpool and Manchester areas. They were also charged with maintaining contact with Pilkington workers in other parts of the country, and with dealing with incoming telephone calls from the press, television, and people from all over Britain offering support.

The other main administrative affairs were dealt with by the treasurer and two assistants. Their task was first to build up a list of contacts and then to send them letters appealing for financial assistance. Having done this the main job was sending out receipts and letters of thanks. (The RFSC was scrupulous in its money matters; accounts were audited weekly, and the treasurer was noted for his 'carefulness': as the resident wit put it: 'If he was in charge of family planning every woman in the country would be pregnant.') The treasurer was also the RFSC's propagandist, was responsible for issuing press statements, for the preparation of strike bulletins, and for stating the RFSC's position to the Court of Inquiry.

THE RANK AND FILE STRIKE COMMITTEE (1)

Picketing

Routine picketing was always a problem. It had, to be effective, to be carried on for twenty-four hours a day and not many people were prepared to work shifts during a strike. It is impossible for us to say just how many people were active pickets. The RFSC estimated that they had about 200 hard-core pickets, with another 200 or so who could be called out at short notice. However, many more could be relied upon to turn up when there was an indication of strike-breaking on any appreciable scale. We would guess that probably as many as 1,000 to 1,500 were engaged in picketing at one time or another.

Responsibility for the pickets at each gate tended to rest with one man who was a member of the RFSC and who would subsequently report back to the committee meeting. Apart from those times when there were attempts to organise mass returns to work, the pickets' main job was to stop lorries going into the factories. Drivers would be questioned – and sometimes asked to produce a union card. If the lorries were taking in supplies for Pilkingtons the drivers would be told of the strike and asked not to deliver. No record was ever kept of the successful approaches made, but from what we heard the pickets met with a favourable response more often than not. A lot depended of course on the RFSC's success in contacting the shop stewards in the transport depots of the firms making the deliveries – several RFSC delegations were sent on these errands and made fruitful contacts as a result.

The RFSC was aware that its main problem with the pickets was the maintenance of morale, particularly of those who stayed on throughout the night. One of the reasons for the exhaustion of so many RFSC members, especially the chairman, was that they spent so much time at night and in the early hours going round the different pickets, talking and listening to their staunchest supporters. But the pickets also drew support from other quarters. Several of the factories were hemmed in by densely packed terraced housing, and women from these streets frequently supplied the men with tea – with such frequency in one area that the RFSC felt obliged to reimburse the women. In another area a friendly publican used to turn up each night with sandwiches and other well-disposed people who were at work would take a picket into a pub for a drink or pass out a few bottles of beer. In another case the owners of a social club handed out the odd bottle of whiskey, and on one occasion gave the proceeds of a Saturday night dance to the RFSC. Relationships with the police and the Pilkington security men seemed to have been amicable when things were quiet: they talked and drank tea with the pickets. No doubt the odd bobby was grateful for the warmth of a brazier on a cold night and a bit of company to relieve the tedium of the night patrol.

'Blackings'

'Blacking' simply means persuading workers in other industries not to handle the goods or supplies of the firm which is in dispute: it is an operation of primary importance when a strike is not completely solid. In this particular strike blackings were of fairly minor importance and probably only constituted a minor irritant to Pilkingtons. However, if other of the RFSC's attempts at getting goods blacked had been successful there would have been a far more serious disruption of the motor industry.

There were two commodities without which Pilkingtons would have been in serious trouble: hydrogen gas and oil. Both were needed to keep the glass furnaces at a suitable temperature so that glass production could be resumed within a few days of a return to work. If the tanks had been allowed to cool, the contraction of the materials out of which the tanks were constructed would have been so serious that it could have taken up to three weeks to get them on line again.

In the fifth week the RFSC decided to try and black gas supplies at the Cowley Hill works. They contacted transport shop stewards at British Oxygen plants in St Helens,

THE RANK AND FILE STRIKE COMMITTEE (1)

Widnes, and Swinton, but found that it was not delivered by road: it was pumped through to Cowley Hill from a plant in St Helens. They did manage to persuade men at this plant to refuse pumping operations but were thwarted by the staff who took over the job.

Two other blacking attempts were made with varying success. Pilkington transport drivers refused to handle any glass until there was a full return to work, and the local AEF branch passed a resolution to the effect that their men in Pilkingtons would refuse to do the work necessary to get production going until there was a full return to work. A previous attempt by AEF shop stewards to get all the maintenance men out on strike in support had failed.

The RFSC considered that one of the most damaging ways of getting at Pilkingtons would be through the motor industry: the Pilkington subsidiary, Triplex, produced the bulk of toughened glass for car windows. They thought that if they could stop the car industry, the motor manufacturers would put the 'screws' on Pilkingtons. The RFSC discovered from the press and through shop stewards in the motor industry that some car-makers, notably Ford, had found alternative suppliers on the Continent. Since much of the imported glass was being flown into the country they attempted to contact cargo handlers at some of the airports and persuade them to black the glass. One delegation went to Liverpool airport where, so one press story had it, there were fifteen planes a day flying in glass. The handlers agreed to black the glass but were subsequently threatened with dismissal if they maintained the ban. The upshot was that only the night shift continued its ban. The RFSC also received a telegram from the handlers at Southend airport to the effect that all foreign glass was black. Large quantities of glass were also coming into the country through the roll-on-roll-off container quays at Felixstowe, but no attempts were made to contact the dockers there.

Overall, the blacking operations were not successful – partly because the RFSC's heart was not really in the job, partly because they were too inexperienced to know how to go about it properly, but mainly because their organisation was already over-taxed with more pressing problems closer to home.

Mass meetings

The RFSC organised twelve mass meetings in all and even Pilkingtons were impressed with the speed with which the RFSC seemed to be able to muster never less than several thousand men. The technique was simple: a couple of cars fitted with loudspeakers, piles of leaflets in Labour clubs, and leave the rest to word-of-mouth. It worked very well in St Helens though it could hardly work in a big city.

The frequency of mass meetings was occasionally a matter of debate in RFSC committee meetings: one school of thought was that meetings should only be held when there was something concrete to report, the other was that meetings were necessary reminders of solidarity and that anyway there was always something to report. The latter were in favour of frequent meetings – and they had their way. Tactics at mass meetings were not usually discussed in committee though there was a debate on one occasion over the way the motion to continue the stoppage should be put. The usual form was: 'All those in favour of continuing the stoppage please show ... all those against please show ...' That section of the RFSC that would have liked to have called the strike off wanted the form to be reversed. The usual form was retained. Tactics for the meeting were usually hurriedly discussed a matter of minutes before the meeting started and speakers just as hurriedly picked. The reason for this general lack of organisation was obvious – the RFSC's ideology was that the meetings had to be democratically conducted, and that meant that recommendations had to come from the floor rather than from the platform. This rarely worked in practice because the RFSC on the one hand didn't have any clear ideas as to how to make it work, and the rank and file on the other hand were just not accustomed to ways of conducting meetings and did not know how to react when offered the chance of participation. The result was that valuable opportunities were missed to experiment with forms of mass participation.

It could *not* be said that mass meetings were used either to sway people or were stage-managed. The meetings were generally short – about twenty minutes or so, and the RFSC did not have any orators, any spell-binders: there was no thunderous rhetoric, no lightning wit. Just ordinary men, unused to such exposure, fumbling for words. The one fluent speaker seemed so hopelessly out of place in such company that he positively jarred. We have said that mass meetings were not 'stage-managed'. This is not quite true, for two of them were, though perhaps 'stage-managed' is too strong a word. At one meeting there was a mass resignation from the GMWU – the RFSC having duplicated appropriate forms in advance. But of course the RFSC would never have distributed them if they had not been confident of getting a favourable response. The other meeting was not so much stage-managed as a piece of spontaneous theatre – the chairman, angered like other RFSC members by allegations of 'subversive elements', called out members of his committee one by one and said to the crowd: 'You know these men – are they reds, are they Nazis?' The crowd roared back: 'No, no.'

The fact was that the RFSC mass meetings were very orderly and by no stretch of any but the most hopelessly biased imagination could have been said to be occasions of great emotion or 'mass hysteria'.

Delegations

There were a large number of delegations, and RFSC members found themselves travelling around the country and doing things completely outside their previous experience – from addressing mass meetings in Glasgow to talking with a government minister in the Ministry of Social Security; from arguing with national officials of the GMWU at the TUC to picketing the gates of Pilkingtons'

DELEGATIONS

Pontypool factory. The RFSC, as they never tired of saying, learned a lot. In the space of a few weeks they had an intensive course in the politics and strategy of trade unionism. It would be difficult to over-state just how important all this activity was to them. Here were men whose social and political horizons had barely extended beyond the frontiers of St Helens before the strike, meeting people and doing things which revealed to them the magnitude of their enterprise. It brought home to them very forcibly that a glassworkers strike in St Helens was more than a little 'local disturbance'.

*

There were strategic delegations and delegations concerned with garnering financial and moral support. In the first category were those sent to Pilkington factories outside St Helens – either to give support to those still out on strike, or to persuade those who had gone back to work to come back out again. These were invariably unsuccessful – all of the Pilkington plants outside St Helens had returned by 5th May two weeks before the main strike was over, and none of these plants came out again in a full stoppage. One section in the Pontypool plant did strike again, but not at the instigation of the RFSC. (The case of the 'Pontypool men' was to complicate the end of the strike in St Helens for the twenty-eight men involved were subsequently dismissed by Pilkingtons. The St Helens strikers replied to this move by resolving not to return to work until the 'Pontypool men' were reinstated.) Other delegations that could be called strategic involved a visit to the House of Commons and to the Ministry of Social Security in an attempt to get the minister to exercise his special powers and make benefits available to the men and women strikers who were single and in very real hardship. (Several young men were absolutely penniless, sleeping rough in disused brick kilns, and relying on charity for food. An RFSC member once gave one of these men the last of his own money: three shillings.)

THE RANK AND FILE STRIKE COMMITTEE (I)

It was largely because of these hardship cases that the RFSC devoted so much of its energy in the last few weeks stumping the SW Lancashire area trying to get financial help. These delegations were much more successful – if the sums were not always large, they rarely went away empty-handed.

'Outside influences'

Many allegations were made during the strike that it had been infiltrated by 'subversive elements'. These we can categorically state were completely without foundation. We were closer to the RFSC than anyone else from outside – how close can be indicated by the fact that we were often mistaken for members of the RFSC and that we were often approached by journalists after inside information. Many were the free pints we had from journalists who were hoping to loosen our tongues. We missed very little of what was going on, and we are quite sure that we missed nothing of importance.

Other outsiders were at work in attempts to prove or disprove the stories of subversive infiltration: professional 'sleuths' working for *The Times* and *The People* were in town for a couple of weeks but reported negatively to Pilkingtons and to their papers. The Special Branch of the police, although unsuccessful in infiltrating the RFSC, also reported negatively to Pilkingtons. A copy of the report did not, of course, go to the RFSC. One Special Branch man, hopefully decked-out as a 'leftie' – long hair and beard, etc. – put in appearances on the picket lines but the only information he received was a thump on the jaw. Another, posing as a press cameraman, was scared off by a genuine cameraman who tried to photograph him! In short, the most effective 'subversives' in town were on the side of established authority.

Representatives of the various left-wing groups – mainly from the Socialist Labour League and the International Socialists – were naturally around, and we spent many an

'OUTSIDE INFLUENCES'

intriguing evening watching them trying to approach RFSC members. RFSC responses ranged from polite attention and a laugh afterwards to a blunt 'piss off'. Only one outside group successfully got inside an RFSC member's house – Moral Rearmament (MRA), an organisation well known for its right-wing views! The chairman of the RFSC arrived home one night to find two MRA men showing his wife a film about how a Brazilian unofficial strike leader came to find that employers and trade union leaders were really quite nice people. He got the message – and didn't like it. Just as other RFSC members got left-wing messages and didn't like them either. As one member put it:

'There were some people I got to know during the strike who used to irritate me. There was one man, X, who used to speak down to us and he really got my back up. I know he wanted to help us but the way he spoke narked me, and a lot of other people felt like that too.'

The RFSC was very selective in whom it listened to. If the representatives of the left-wing groups got scant attention, the shop stewards from other firms were regarded with respect – awe even in some cases. It was after a meeting with some shop stewards from Fords in Liverpool that RFSC members started to talk about the strike being an education. It was also after that meeting – thirty-six hours later – that the RFSC organised a mass resignation from the GMWU.

The Ford's shop stewards told the RFSC about the trouble they had had with the GMWU some twelve months previously and how 2,500 men had left the GMWU, and joined the T&GWU and AEF. The next day the RFSC duplicated forms instructing Pilkingtons to stop deducting union dues from their wages when they returned to work. The following day, Sunday, 10th May, Ford's stewards spoke at a mass meeting telling the crowd of their experiences with the GMWU, and when the RFSC chairman called for a vote in favour of resigning from the GMWU the air was filled with white paper. From that time on until a month or so after the strike most members

177

of the RFSC thought they would have little difficulty in joining the T&GWU though there was still a sizeable number of the RFSC who were prepared to give the GMWU another chance.

Other outside influences were at work, but they were of a completely different order: hundreds of letters offering moral and financial support from all over Britain. They came from other workers, branches of other unions, branches of the GMWU, shop stewards committees, from constituency Labour Parties, from Trades Councils, from students, from housewives, from old age pensioners, a few even from managers, and one remarkable letter from a senior company executive. A selection of these letters appears in the Appendix.

The RFSC received in all, through donations, approximately £3,500. It came in sums ranging from 6/4d from the wife of a retired Liverpool docker, a pound from an eighty-four-year-old Pilkington pensioner, to £800 from workers on a power station construction site, £400 from another large construction site, £200 from Ford workers at Liverpool, and £500 from British Leyland workers at Leyland, Lancs. Liverpool University students managed £28, and students from many other colleges and universities sent contributions ranging from £2 to £20.

Donations poured in at such a rate in the last week that, together with firm promises they had received of regular weekly amounts of £400 in two cases, the RFSC could, had the strike continued, have started to make hardship allowances to especially needy cases. Indeed, the RFSC had every intention of doing so for in that week they were searching for premises suitable for the distribution of money. The flow of money was helped considerably by unsolicited advertisements for funds appearing in such papers as the *Morning Star*, *Socialist Worker*, and *Black Dwarf*. This was the nearest the left-wing groups came to influencing the course of the strike.

Thanks largely to the parsimony of the RFSC treasurer the bulk of the money received was still in the bank at the

'OUTSIDE INFLUENCES'

end of the strike and this helped considerably when the RFSC, at the beginning of July 1970, set up their breakaway union – the Glass and General Workers Union. The total cost to the RFSC of running the strike was somewhere in the region of £750 – most of the money being spent on train fares and petrol for the various delegations; the remainder went on postages, duplicating expenditures, and payments of £25 to the 'Pontypool lads', and £25 to a man with five children who was widowed during the strike.

The RFSC received a great deal of encouragement from the letters and money sent to them. The support also came as a surprise, particularly to those who had not previously been active trade unionists. It had never occurred to them that solidarity was anything other than a local phenomenon: their resolve was hardened when they found that the postmark of just about every British city of any size was represented in the morning mail. They were amazed and delighted to find they had so many friends.

NOTE

The information on which this and the next chapter is based was collected in a variety of ways. For the last three weeks of the strike we attended all RFSC meetings – formal and informal, public and private. We slept in their homes, had meals with their families, rode in their cars. We sat in TV studios with them, accompanied them on delegations, attended press conferences with them and shared their train compartments. Wherever they went and whatever they did, we were there with our notebooks.

In the two weeks immediately following the strike we interviewed twenty-five members in their homes. The interviews lasted on average about three hours – about two hours of conversation was followed by one hour of questioning recorded on tape. Our respondents needed no encouragement to talk freely.

We enjoyed, we suspect, a unique research experience.

Chapter Seven

THE RANK AND FILE STRIKE COMMITTEE (II)

Those who are disposed to think of the RFSC as a collection of hardened 'class warriors' are shortly to be surprised, or disappointed, depending upon their point of view.

If the RFSC were obviously militants in the *strike* situation, it became very clear to us in the course of our interviews that most of them did not ordinarily see themselves involved in a perpetual struggle with management. There were a few – four out of twenty-five – who did see relationships in this light, and this group accounted for *most* of those who played prominent parts in the RFSC. The other members however, while displaying a *latent* tendency to think in terms of struggle, were prepared to think of everyday activity on the shop-floor as co-operative. People in this group almost invariably held contradictory views and this made it difficult to classify them. Yet it will become clear as we proceed that the contradictions in attitudes were expressed in quite different ways. We have therefore divided the non-militant majority into two groups.

There are the fatalists: 'We will always be on the losing side but sometimes we shall have to fight to hold on to what we've got.' And there are the reformers: 'OK, so what's good for the managers is not always good for us, but there is always something to be done to improve our lot.' (It need hardly be said that none of the RFSC actually expressed themselves in these words.)

The responses to some of our questions exemplify these three different standpoints:

THE MILITANT
What was the normal state of labour–management relations?
'Out of date, out of touch. It's management's policy to keep men against men, department against department with the

piece-work system. Management is arrogant – it's the only way they know of talking to the working class. They have this attitude of – "We know what's good for you. We'll stand no argument".'

What do you think about the balance of power between the management and the shop-floor?

'It's not right as it is, but we don't want men on the board of directors because they'll probably end up as bent as the bosses. Too much participation with the bosses and you end up by losing touch with the reality of the shop-floor.'

Do you think Pilkingtons control St Helens?

'Pilks may not control the Council, but when Harry shouts they jump. He has the pull where it counts. They've talked about subversive elements in this strike, but they've had subversive elements in for a 100 years. They've infiltrated every organisation – right down to charities and raffles.

'They've not got their men in everywhere, but all they need to do is to give a nod and a wink. They've conditioned people to think the way they do – it's a way of life for some people, is Pilks. The Gala's, the OAP's treats, the way they introduced family allowances years ago – they've had a cast iron grip on people's minds.'

What qualities are needed to make a good shop steward?

'His first responsibility is to his members, but I think too many toe the union line and ignore the shop-floor. There are others who get too cosy with management – sit too much in the office drinking tea. You've got to keep your distance.'

THE FATALIST

What was the normal state of labour–management relations?

'Before the strike at X it wasn't too bad. The staff were very friendly and the atmosphere used to be very happy. The works manager in X, Mr P, and the personnel manager, Mr Q, were all right, I never had any complaints. If ever we had reason to complain they were fair when we went in. There was very little trouble.'

What do you think of the Pilkingtons?

'I think the whole family is domineering. They treat the working man as if they are doing him a favour by keeping him in work; this is the attitude I get from them. It's always been like that.'

What do you think about the balance of power between management and the shop-floor?

'The balance of power is all on the management's side – it always has been and always will be.'

What qualities are needed to make a good shop steward?

'Plenty of fighting spirit and commonsense. A sense of brotherhood with a little bit of Christianity in it. He should be fair-minded, strong-minded, and prepared to go to any lengths for his men. He should be prepared to fight on for evermore against any type of employer.'

What do you think of Lord Harry?

'I don't think much of the way he's treated St Helens. He's been the Lord and Master for so many years that you end up thinking he always will be. He's done it for so long now that he'll never be able to get away from it.'

THE REFORMER

What was the normal state of labour–management relations?

'Inadequate, because the shop-floor is never allowed to participate fully in decision-making which affects them. The works manager is aloof – he doesn't seem to have any deep interest in the employees as persons. His main interest was in getting up production; complaints, grievances, were a burden, a chore he didn't like doing. Sometimes they are arrogant. They have an air of superiority which seems to say "you're just a peasant". Some have been genuine and tried to be fair, but their hands have been tied. They haven't had sufficient autonomy to do things they knew needed doing.'

What do you think of Lord Harry?

'I think he's been wrongly informed and completely out of touch. I wouldn't blame him wholly for what's gone on. He may of course have condoned the actions of the people responsible, I don't know. But he has to accept the responsibility as chairman. He could have shown more leadership – if he had wanted to end the strike at an earlier stage he could have said, "to hell with normal procedures" and met the rank and file committee.'

What do you think about the balance of power between management and shop-floor?

'We must push for more participation in decision making. Workers should be allowed to express their views all the way up the line to the board of directors. We must say, "This is as much our firm as it is yours. We put everything into it. The decisions which you put into effect, affect us." We may not be able to achieve this under the existing capitalistic set-up but we must try just the same because the firm doesn't just belong to the shareholders. It belongs to everyone who puts any effort into it.'

What qualities are needed to make a good shop steward?

'He must be a sympathetic listener. He should be able to discuss grievances with the people he represents and be able to express their grievances. He should be able to demand and to prove to management that he has the full support of those he represents.'

*

What distinguishes the militant from the others is that it is only he who consistently gives answers in terms of struggle; it is only he who sees every relationship in terms of us and them. The militant is a theorist for he sees relationships in terms of categories. The fatalists and reformers by comparison tend to personalise *relations* and focus their attention on the relations themselves: they see the *content* of relationships rather than an organising principle lying behind them.

But this is not to say that the fatalists and reformers do not see themselves as involved in struggle – the quotations show very clearly that they do. The difference is that whereas the militants *consciously* look at situations as examples of a power struggle, the others do so *sub*consciously. That is to say that for the fatalist and the reformer struggle is at the *back* of many of the things they say. The differences between these types then are not as clear cut as they may seem, particularly when it is borne in mind that on some points at least there is a fair degree of overlap. The types indicate only broad attitudes of mind.

In a situation such as a strike, where it becomes plainly obvious to everyone that a struggle is involved, the differences become submerged. The militants, the fatalists, the reformers, line up together because the crisis is transparently one in which sides have to be taken. Nevertheless the differences between these types were to surface at different times during the strike, especially when tactics were being discussed, but it was not until the last week that they became clearly evident. The fact that differences did arise, and that they could be hung on the doors of the people who broadly fell into the three types, helps towards an understanding of how the strike ended when it did.

An element of surprise

About one third of the RFSC expressed surprise that the strike should have occurred:

'I was surprised by the way it spread because I have always looked on St Helens people as a lot of sheep who would lie down to lots of things. But I felt great when I saw how solid people seemed to be – I was up a tree! It's a great feeling to see people solid like that – when they are solid you can climb mountains.'

But not everyone else was that enthusiastic:

'I didn't want a strike because of my home situation. When you're young and married with three kids you don't want to lose any money. But I knew that some sort of action was going to be necessary so when the strike started I thought we had better take advantage of the situation.'

But if some were surprised at the way it started and spread, most had been half expecting something to happen for some time:

'I had expected a strike for a long time. With being on the branch committee you get called to emergency meetings, and I'd been called to three in next to no time. When you get called to a branch committee meeting and they have never called one for twelve months, and then you go to three in two months, you know there is something wrong somewhere.'

Another RFSC member found it easier to understand in retrospect:

'We knew it was coming. We didn't know when, but we knew. The reason it hasn't happened before was lack of shop-floor organisation. But at Triplex where I worked we had organised a works committee and we'd had feelers out in other factories – where we'd found that their grievances were our grievances as well. Where we thought we were fighting it alone at Triplex we found we weren't. Everyone else was at the same game and the same aim. So I think it just exploded all together once the shop stewards found they weren't on their own.'

At least one member of the RFSC was *relieved* at the strike for his section of twelve men from the Fibreglass plant had been out on strike for two days beforehand.

A 'powder keg'?

All members of the RFSC thought there was a general atmosphere of discontent over such things as complicated systems of wage payment, the smallness of previous pay increases, the lack of speed with which relatively trivial issues were settled, and the ponderousness of the Joint Industrial Council. But some members, because they had worked in departments which had been involved in short-lived disputes of one sort or another, thought Pilkingtons was ripe for an explosion. Short stoppages lasting anything from an hour to a couple of days, overtime bans and 'work to rules' of similarly short duration if not widespread, had been common enough in some departments:

'We'd had about four disputes in our department – refusals to do a job. The men had threatened to walk out several times but never did because management always came and said: "Leave it with us." So we'd start back to work. Nothing happened of course so we had another walk-off, and management would come in and say: "This time we'll definitely do something." But they never did. We'd had a walk-off about a fortnight before the strike.'

Disputes of a similar nature and duration had been something of a commonplace in the Flat Drawn department of Sheet Works where the strike started, but at Triplex there had been stoppages lasting for a few days:

'We'd been out twice before this one – we did a two days and a four day strike in my department. Negotiations had been going on for twelve months over a bonus issue but nothing happened so we thought we'd show them that we meant what we said. Six months ago was the first strike – that was for two days, and then we came out for four days about six weeks before this one was started.'

All in all, although we were unable to get a reliable count of the actual number of disputes that had taken place, we were clear in our own minds that in at least four of the six plants there had been an increasing number of short-lived disputes of one sort or another. We are also clear that because most of them rarely lasted the life of a shift and did not disrupt production, they did not come to the attention of the head office industrial relations staff.

Apart from all these small incidents RFSC members were unanimous that the main reasons for the strike were:

'The low basic rate of pay, a difficult wage structure, a complicated bonus system, the playing off of one group of workers against another by giving differential increases all led to tension.'

'It has built up over the years. The Joint Industrial Council is just an organisation to shove things through without the shop-floor knowing about it. This distrust has built up over the years.'

'New blood. They've employed men from outside who have worked in other industries, they've seen what's wrong, voiced their opinions, and been listened to by the St Helens people. People who've worked there all their lives haven't seen things *because* they've worked there all their lives.'

'People from outside St Helens drifted in – this was one of the finest things that has ever happened at Pilks. This type of lad just wasn't prepared to sit down and wait – he wanted an answer. This coupled with the fact that the old-stagers of St Helens had been boiling for years over injustices in the wages structure. The ones from outside probably started it, but there was fodder waiting to be put on the fire.'

'Everything was bubbling away and sooner or later it had to explode. It wasn't just the young workers, in fact at Cowley Hill it was some of Pilkingtons' long-serving and 'most trusted servants' who were most militant. Their children had grown up, they'd paid off their mortgages, and it was a chance for them to make up for the suffering they had endured in the past.'

Given this generalised air of discontent, and given that there had been isolated outbursts of militancy, it becomes easy to understand how sufficient men in each works could be mobilised for action. Action which could very rapidly focus the *generalised* discontent into a very specific issue.

The question of paternalism

By concentrating on purely economic issues such as bonus payments, wages structures, negotiating procedures etc., the fact is often missed that a factory is as much a *social* system as a producer of wealth. This is another way of saying the obvious – that economic questions are set in a social context. The reader need only consider that when a man works in a factory he gives up his freedom as well as his labour in return for a wage. It would therefore be disastrous to set to one side as of no account the attitudes and values of management on the one side and workers on the other.

We did find amongst the RFSC a strong note of protest that they were often treated as less than human – the words 'serf' and 'peasant' were commonly used. This protest was evident amongst militants as well as amongst fatalists and reformers, although it was the fatalists and the reformers who felt most strongly that Pilks had violated the family tradition:

'I've met Lord Harry a few times and I think he'a a decent bloke. He doesn't know what has gone on below and I don't think they want to know until we get a crisis and everyone wants to know how it has happened. It's not so much been the family as the organisation that's been wrong.'

'I don't think that Lord Harry know's what's going on. It's the organisation as runs it. They're codding him that everything is running smoothly and the union is codding management likewise.'

'I think he's taking the can for us. It's my honest opinion that it's not him. He's only one vote on the board. I heard that when float was licensed to Russia Lord Harry wanted to give all the workers £25 apiece. When float was licensed to the Yanks we all got £10 each, but this time he was outvoted.'

'Well, I think Lord Harry is the same as Elvis Presley – he doesn't get amongst his fans. He just doesn't come and visit us often enough. He's an enigma, a name. We never see him – he could be dead for all we know.'

The feeling that the family was basically well-intentioned was a view mainly held by the fatalists for it was this group that found no difficulty in distinguishing between the family and the organisation, and holding the organisation at fault. But on the other hand virtually all held that top management was completely out of touch with the feelings of the shop-floor workers. The RFSC member who said: 'The top management haven't got a bloody clue of what's going on. But I suppose if they did it wouldn't help. In fact I think they'd be a bloody sight tighter if they did know,' was almost on his own because there was the strong implication amongst all the rest – bar the militants – that if top management *had* been in touch things would have been much better. If only one man said: 'I suppose if top management knew that our manager was penny-pinching and hindering production

THE QUESTION OF PATERNALISM

they would definitely take a hand in it,' the others had the view at the back of their minds that the field-marshals were OK, but the platoon commanders and the sergeants stopped any bad news filtering through to the top. This is an attitude of mind not dissimilar to that of rebellious peasants in Tsarist Russia:

'The peasants justify their rebellion on the ground that the Tsar's authority has been abused ... Now, the claims made on behalf of the Tsar's authority has always been that he is a benevolent father who looks out for the welfare of his people. Accordingly, the rebels appeal to the official creed of the Tsarist order, when they interpret their massive deprivations as evidence that the Tsar's authority has been abused. For a rightful Tsar would protect his people against oppression; he would safeguard the just claims of even the lowliest peasant.'[1]

In spite of the remarkable resemblance between this and the views so often expressed by RFSC members, the analogy should not be pushed too far. In the first instance the RFSC did not *justify* its revolt with reference to abused authority, and in the second instance their views of authority relations were a mixture of the traditional (deference to authority) and the modern (distrust of authority), and accordingly not a little ambivalent. Thus while there was an affirmation of the traditional order in Pilkingtons, there was a good deal of hostility towards the way Pilkingtons were seen as dominating St Helens:

'Going back, they were the major industry in the town apart from the mines. We know of instances where they have stopped other industry from coming in. They have dominated the labour force of St Helens. And having the whip hand over the labour force, they have dictated the wages structure of the town. I'm sure they have influenced politics as well – the council always paid at least lip service to Pilks because they are so powerful.'

One man said of Lord Harry: 'I think he's been interested

1. R. Bendix, *Nation-Building and Citizenship*, Doubleday & Co., New York, 1969, p 55.

for years – I've actually seen him on his bike. For me, I think he's really interested in St Helens, but he's got out of touch in the works. I don't think he puts enough time in walking around the works.' But when we asked him if he thought Pilkingtons controlled St Helens, he said: 'They have a big say in the town. Pilkingtons have convinced people that they are the saviour and will look after them.' If few other RFSC members displayed such marked contradictions, ambivalence was nevertheless almost invariably present.

Civil War in the GMWU

By the third week of the strike Pilkingtons receded into the background – the firm was beginning to wonder if the strike had anything to do with them at all. As regards the union the RFSC had no feelings of ambivalence whatever – just complete, uncomplicated feelings of betrayal. The sense of outrage was so deeply felt, the sense of injustice so acute, that it was hard for RFSC members to express themselves. When one member said: 'Words fail me,' in answer to our questions about the union, he was so choked with emotion that there seemed little future in pursuing the question.

While there had been a good deal of suspicion of the union from the outset most RFSC members felt the situation was such that the union would have to do something. And of course the GMWU did, for it had after all negotiated a £3 a week increase. Most RFSC members were taken completely by surprise when two mass meetings of strikers turned down the £3 offer but were amazed and horrified when the GMWU refused to accept these decisions and told the men to go back to work. They were not amused either by the way the GMWU and Pilkingtons worked so closely together after that point in trying to get a return to work.

The man who said: 'I didn't expect anything from the union at the start. They had no leaders and no organisation: it was just a firm's union. They were hand-in-glove with the firm and we've known this for years,' was an exception. The dominant view was: 'I honestly thought the union would take up our demands because all the stewards were 100% behind the strike and so was the branch and the branch secretary.' As another man said: 'When the branch secretary said that it was official at branch level I had every hope that it would be made official at national level within a week and that the whole thing would have been settled within a fortnight.'

These were men, it must be remembered, with no previous experience of strikes and with very little knowledge of the politics of trade unionism in general and of their own union in particular. For most of them the GMWU was either their shop steward or the local branch. None of them had ever aspired to office in the union. They had never had occasion to consider seriously their union's policies or its structure. As with Pilkingtons, there had never previously been any crisis to focus the vague and general dissatisfaction.

But once the union had turned its back on the strikers they started to look at it a little more closely. If one man summed up the general view by saying: 'I feel really bitter about this union. In seven weeks it hasn't done one thing to help its members,' others were more specific:

'The real thing that came home to me was that no matter what our immediate officials, such as the branch secretary, the area organiser, or the district secretary did, there was nothing I could do about it because they are appointed and not elected. What this means is that the National Executive can say that these people stay in office no matter what the people in St Helens might think.'

'The union was despicable. I don't think for a moment that they give a hang for the men at Pilkingtons: they are only concerned with the structure of the union, with getting as many members as possible by having closed shop agreements. The union believes in no unofficial strikes, low wages, and keeping the men down. I don't look upon the union and Pilks as two separate bodies – they are hand in

THE RANK AND FILE STRIKE COMMITTEE (II)

glove. They just wanted to break the strike and get the men back to work. They tried it any way possible; they lied, they conned. It was just despicable.'

'I think it handled the strike completely incompetently. It refused to make the strike official. It refused to see the seriousness of the situation. It wanted to control and to dictate to the membership rather than listen to the membership's views.'

What especially hurt a number of RFSC members, particularly those who had been shop stewards, was that prior to the strike they had often defended the union against criticism from the shop-floor. They had said, as all shop stewards must say almost daily: '*You* are the union – the union is as strong as its members.' And then they found themselves in a strike discovering that the members were *not* the union, and that the members were giving the union a strength the officials did plainly not want.

Criticism – a mild word for what was actually voiced – centred around four main areas: the failure of the GMWU to declare the strike official; GMWU tactics; its collaboration with Pilkingtons; and its failure to be democratic.

On the first Monday of the strike the branch secretary – a full-time official of the union – had told a mass meeting that the strike was 'official at branch level'. The recommendation to the National Executive Committee of the GMWU from the St Helens branch that the strike be made official at national level was rejected. This was incomprehensible to the RFSC. There were 8,000-odd GMWU members out on strike and continually voting to stay out on strike – if they were the union, as they had so continually been told, why weren't *their* officials carrying out the wishes of the membership? They thought trade unions were democratic, and democratic meant going by what the majority said.

If in the early stages criticism of the GMWU's lack of democracy centred around its failure to make the strike official, it subsequently became more sophisticated as RFSC members started to look at the union rule book

CIVIL WAR IN THE GMWU

more closely and to learn something about trade unions from their contacts with shop stewards in other firms. They also learnt something from the pamphlet *GMWU – SCAB UNION*, published by the London-based syndicalist group, Solidarity. Two hundred of these pamphlets were given away in St Helens. (The GMWU, some months before the strike, had taken legal advice as to whether or not to sue Solidarity but did not eventually pursue the question of a court action.)

The influence of shop stewards from other firms was undoubtedly considerable:

'It was only by coming into contact with people such as Fords shop stewards, people in the AEF, construction site men, members of the Labour Party, that I suddenly realised I had to look up my own rule book. I had acted as a shop steward for the GMWU but all of a sudden I realised that there are other unions. I was very naive as far as other unions were concerned. I thought all unions were the same – it was only then that I realised that there was something wrong with the set-up, something definitely wrong, and that something had to be done about it.'

GMWU tactics served only to inflame the RFSC: tactics seen as being aimed at discrediting the RFSC and at securing a return to work through collaboration with Pilkingtons. They were not altogether impressed with Pilkingtons' tactics either, but Pilkingtons were a much more minor source of affront. They were, after all, employers, but the union was in some way 'ours'.

A number of things the GMWU did, or was alleged to have done, fortified the RFSC's feelings of being a small group of sincere, honest men, in a world otherwise peopled by scurrilous liars who would descend to the ultimate depths of treachery and chicanery. They were convinced that several groups of strikers in Pilkington plants outside St Helens had been got to return to work because they had been told by union officials that the St Helens strikers had returned to work. They were appalled when the GMWU arranged a press conference in Liverpool to allege that the strike had been infiltrated by 'subversives'. They were horrified by the union's failure to declare the result of the postal ballot. But most of all they were embittered by the union's persistent attempts to break the strike, not least when at the end of the fourth week the GMWU organised meetings at each of the factory gates. How could the union do all this to them when all they were doing was to repeat the union's original platform of 'no return until we get £25 a week basic pay?'

By the sixth week they were not prepared to see the concurrence of the 'parsons' poll' and the payment of hardship money by the GMWU as coincidental. They dismissed the hardship payments as a bribe to get people to vote to return to work in the poll which followed close on the heels of the distribution of money. Hadn't the union said it was paying the money because men had been prevented from returning to work? How then could it give out the money indiscriminately? How could its reason be true when both prevented and preventers were paid alike?

Come the 'parsons' poll' and the RFSC were prepared to believe anything of the GMWU. They alleged that the priests had been led by the nose and that the ballot had been rigged – the RFSC found two men who said they had been signed into union membership only a few days before the ballot, another who had been refused hardship pay but allowed a vote, and another who joined the union the day before the ballot, collected his cards from Pilkingtons on the same day, and voted on the next.

Collaboration with Pilkingtons stuck in their craw too. They knew that it was with Pilkingtons' assistance that the St Helens Rugby League ground had been made available to the GMWU for its mass meeting on 13th April; they knew that Pilkingtons had addressed envelopes for the union's postal ballot; they knew that Pilkingtons had provided the union with loudspeaker equipment for its factory gate meetings on the weekend beginning 1st May, they knew that the union branch register was out of date and that the union had to rely on Pilkingtons for an up-to-date list of

PILKINGTONS, THE POLICE, AND THE PRESS

voters for the 'parsons' poll', they knew that a substantial part of Pilkington's letter to its employees, dated 12th May, had been written by the union; they knew that after the rejection of the £3 offer there were frequent meetings between representatives of the union and Pilkingtons and that Pilkingtons offered advice. And the fact that they knew so much of all this only served to strengthen their conviction that the union was in 'Pilkies pocket'. (They knew so much because they had a very good intelligence service: the RFSC had sympathisers in unlikely places.)

Pilkingtons, the police, and the press

There were several things either said or done by Pilkingtons that reinforced the RFSC's view of itself as the only honest men in the situation. Over the weekend of 3rd–4th May, when the GMWU was attempting to organise a return to work, the management at one of the plants – without the knowledge of the Pilkington strike executive – sent people round to the houses of some women employees telling them that if they did not return to work they would be liable for dismissal. The RFSC thought this was the lowest trick in the book, and prompted a remark from the RFSC chairman that the RFSC had behaved like gentlemen up to then, but if Pilks wanted to play it dirty they were prepared to take the gloves off too: it was following from this incident that the RFSC tried to stop the supplies of gas going to the Cowley Hill plant.

It was at this time too that Pilkingtons extended the £3 pay rise to all other employees – office staff as well as maintenance men. This took the maintenance men by surprise for they were not accustomed to getting unsolicited pay rises. It took the RFSC by surprise too – they construed it as an attempt by Pilkingtons to turn other people against them.

During the first attempt at a return to work Pilkingtons gave out figures of the number returning which the RFSC regarded as grossly exaggerated – the RFSC thought that

THE RANK AND FILE STRIKE COMMITTEE (II)

their estimates which were 50% of those given out by Pilkingtons were nearer the truth. They were convinced that Pilkingtons deliberately exaggerated so as to encourage more people to go back to work. Exactly the same allegations were made about Pilkingtons in the last week of the strike when there was a further attempt to get a return to work. It was at this time that a new dimension was added to the strike – the RFSC protests that the police were 'Pilkies private army'.

This protest was based on their belief that the Chief Constable for Lancashire had dined with Lord Pilkington on Sunday, 17th May, and that Pilkingtons had informed their employees in a letter sent out on that same day that there would be ample police protection for those who wanted to return to work. Ample protection there was – bus-loads of policemen and mounted police were drafted in from other Lancashire towns. A total of twenty-six arrests were made that week. The RFSC was very angry about the police, alleging provocation of violence, unnecessary force, and downright brutality. They were however to exonerate the local police who they thought had behaved very well:

'I thought the St Helens police were very good, but the outsiders were militiamen, thugs, and brutes. I know now why the students "call" the police in demonstrations – I'll not blame the students again.' But many others were not so interested in discriminating:

'The police were definitely Pilkies private army and you can print that in ten feet letters. They sided with management by reducing the size of the pickets and as good as saying: "We are breaking your strike."'

'I would say without a doubt that the police were siding with Pilkingtons. There was obviously some plan drawn up with Pilkingtons to break this strike with police intervention.'

The RFSC were very clear in their own minds about the role of the police. They argued that even a mental defective could see that the police were being used to break the strike, for if they hadn't been there no one would have gone through the picket lines. Thus the fact that they were there to escort people into the works, obviously meant that the police were being used to break the strike. And since the police were there in such strength at the request of Pilkingtons, it could only mean that the police were effectively, if not literally, 'Pilkies private army'.

The RFSC was no more flattering about the press, though the 'heavies' like *The Times*, *The Guardian*, *The Financial Times*, *The Daily Telegraph*, and the 'posh' Sundays were almost universally exempted from criticism:

'*The Sun*, the *Daily Mirror*, and *Daily Express* are not worth the paper they are written on. They are papers I used to buy, but I don't any more. Papers I've never bought in my life are *The Financial Times* and the *Observer* – now I do because their people really did try to find out what it was all about.'

As one of the militants put it:

'*The Guardian* and *The Financial Times*, papers the workers don't read, were much better than the others. The popular press doesn't want to educate the workers because the less they know the safer they are. *The Financial Times* and papers like that don't constitute a danger because they are not read by shop-floor workers so they can print exactly what is going on.'

At one of the last mass meetings of the strike the chairman of the RFSC was to say that they had taken on Pilkingtons, the GMWU, the police, and the press. This did indeed sum up the way the RFSC felt – that they had ranged against them some of the most powerful organisations in the country. They were a beleaguered minority which was standing out for justice, fair play, and honesty in a world which seemed not to care for any of those things. They were, in terms of their previous opinions, deeply disillusioned men. If the strike had turned them all into militants, it had turned most of them into fatalists as well. A fatalism which was to be greatly reinforced after the strike by the turn of events which finally crushed them.

THE RANK AND FILE STRIKE COMMITTEE (II)

The magic words – 'The Trades Union Congress'

The RFSC was sick at heart after the result of the 'parsons' poll' had been declared from the Town Hall steps on the evening of Saturday, 16th May. One fifty-four-year-old member of the committee openly wept, and another summed it up for all of them: 'I felt absolutely cut down – as if the bottom had really dropped out of my world.'

By Sunday afternoon they had recovered somewhat – a mass meeting of 3,000-odd people voted to continue the strike – but they never recaptured their confidence. This was the opportunity for those members of the RFSC – the fatalists plus a few reformers – who had lost heart a week or so before to assert themselves. But there were a few militants too, realistic to the end, who realised that if they did not end the strike soon they would be a 'couple of hundred men in the park on our own'.

By the Tuesday (19th May) Vic Feather, General Secretary of the TUC, was offering to mediate between the RFSC and the GMWU, and six telephone conversations followed between Feather and the chairman and vice-chairman of the RFSC. Feather had a consistent and unwavering line: no talks at the TUC without a resumption of work. He held out to the RFSC the *possibility* that agreement would be reached over 'no victimisation', the 'reinstatement of the Pontypool lads', and that some members of the RFSC might be co-opted on to the GMWU negotiating committee with Pilkingtons.

The RFSC attempted to extract from Feather something more concrete. They asked him if he, or a nominee, would visit St Helens and speak to a meeting; they asked if they could have TUC assistance in setting up a new union. They insisted that the people in St Helens would not go back to work on promises because they had been let down too often in the past. They asked for something definite – and in writing. Feather resisted and reiterated his position. The RFSC asked for, and received, a telegram setting out Feather's terms.

MAGIC WORDS – 'TRADES UNION CONGRESS'

The RFSC held a committee meeting that evening to discuss the telegram and decided to put it to a mass meeting the next day – without either recommending acceptance or rejection. As the vice-chairman said: 'The meetings tell *us* what to do.' There were two mass meetings the next day – Wednesday, 20th May – one in the morning, and another in the evening. At both meetings the strikers voted to continue the stoppage.

The mass meeting on the Wednesday morning was chaotic. The chairman of the RFSC, visibly stumbling with fatigue, haltingly read out the telegram which someone else was holding for him (his hands were too shaky for him to hold it himself) but forgot to read out the clause stating that TUC talks were conditional upon a return to work. He then left the platform to telephone Vic Feather as prearranged: Feather restated his position. The chairman returned to the platform after a hurried conference with RFSC members, repeated what Feather had said, and asked for a show of hands in favour of continuing the stoppage – a sea of hands emerged from the ranks.

Almost immediately after the meeting the RFSC realised that many people had thought that they now had TUC backing for the strike – even a BBC news bulletin carried that message. Another mass meeting was arranged for that evening so that the rank and file could be under no misapprehensions regarding the TUC offer. In the evening the RFSC chairman stated the TUC position correctly and asked the meeting whether or not it wanted the RFSC to accept the TUC offer. A voice from the floor called for a continuation of the strike – the motion was put and overwhelmingly carried. Shortly before this, one of the men from Pontypool, a burly ex-miner, tears streaming down his cheeks, pleaded with the crowd: 'Go back for what *you* want and never mind us.' He had judged correctly – the question of the reinstatement of himself and his twenty-seven mates from Pontypool was indeed standing in the way of a vote to return.

On Tuesday the vice-chairman of the RFSC had

THE RANK AND FILE STRIKE COMMITTEE (II)

attempted to contact the works manager of the Sheet Works with a view to persuading Pilkingtons to reinstate the 'Pontypool lads'. On the evening of Wednesday, after the mass meeting, the chairman of the RFSC tried Triplex works management and another RFSC member tried the Cowler Hill works manager with the same end in view. All three failed.

By this time there was a very clear majority of the RFSC in favour of accepting the TUC terms. The committee had already been disheartened by the result of the 'parsons' poll'; they were dismayed by the massive display of strength put on by the police at the factory gates on the Monday and Tuesday of that week and by the number of arrests made; they thought that the resolve of the strikers was weakening and that many more would return the following week; but above all they were overawed by the TUC. Here was the TUC offering to step in, the TUC the greatest guardian in the land of the workers' interests. Surely here at least they would get fair play, here was an opportunity that just could not be passed up.

There were a few who were more realistic and who realised that the TUC General Secretary could really go no further than Lord Cooper, General Secretary of the GMWU, would allow him to go. They realised that Vic Feather in fact had very little power, but the great majority knew nothing of the structure of the TUC. For them the words 'Trades Union Congress' were indeed magic words, words to be conjured with. The realistic minority, aware of this magical aura, wanted to prepare a leaflet explaining the structure and powers of the TUC. But there was an even more realistic minority – of one – who appreciated that this rational response would not work in unsophisticated St Helens. He saw very clearly that such a leaflet would be construed as an attempt to debase the one true coinage – the TUC. He knew that Gods were not to be toppled with leaflets. The leaflet was not prepared.

The picket reports at the RFSC committee meeting on Thursday morning (21st May) were not found encouraging.

'... A BLOODY EDUCATION!'

The number of people picketing had dropped considerably on the previous day, and it was said that the number of people returning to work had increased. This set the scene for a continued debate on the TUC offer.

When the vice-chairman said: 'Where do we go from here now that we've kicked the TUC up the arse? Neither us nor Pilks will budge – we've got to give a little,' he struck a responsive chord amongst a majority of the RFSC. While the debate continued – with no little heat – the chairman was speaking to Feather over the telephone. Feather repeated yet again his offer, but added that the GMWU was willing to meet the RFSC if it would call off the pickets.

The chairman reported back to the meeting, and the RFSC voted by twenty votes to two to call a mass meeting for that evening, and to recommend to it that the TUC offer be accepted. It was, overwhelmingly. The strike was over, but the battle with the GMWU was not.

'By God, this strike has been a bloody education'

The RFSC did not generally mean by this that they had learned a great deal of the detail of everyday trade unionism. What they meant was that they had learned about the politics of conducting a strike. What they had learned in other words was about the reality of power relations:

'I was naive in my outlook towards the union, the company, the press. I always used to think this was a democratic country – but I've been disillusioned. We're just allowed enough democracy and freedom to keep us quiet. I never believed it before, but I believe it now – that this country is run by a minority, the people with the pull, with the money.'

That was a militant speaking, a man who had extended his way of looking at the shop-floor situation to the whole society – he had been politicised even though he did not know it. A large minority of the RFSC however continued to think that Britain was basically a fair, just, and democratic society. Pilkingtons, the police, the press, and the

THE RANK AND FILE STRIKE COMMITTEE (II)

GMWU, were just aberrations in a world that was otherwise still fundamentally reasonable:

'We have gained a vast amount of experience. You could live a lifetime and not gain as much. We have found out what goes on behind industry – the lengths management will go to defeat the working man. There are no tricks too low, no depths they won't fathom. I have had no previous experience of strikes and I didn't know that things like this went on. Pilkingtons descended to the lowest.'

But this was said by a man who also remarked: 'I think, taking the world as a whole, that Britain, in spite of its shortcomings, is one of the best. I think there is still one law for the rich and another for the poor, but if you've got the ability you can really get on.'

For the majority there had been an obvious shift in attitudes from feeling *fairly* optimistic about Britain to feelings of some pessimism:

'Britain gets less and less democratic every day. We are going back to the Dark Ages only it's being done in 1970's fashion. The days when people were persecuted, when they had their hands chopped off for petty reasons. It's now being done in a more up-to-date fashion: you get blacklisted instead so that you can't get a job anywhere. People are getting so that they want to burst out of their skins and take a swipe at somebody for having pressure put on them.'

The movement of attitude was concisely put by a man who is now an ex-member of the Labour Party:

'Britain is unjust and undemocratic. I'd have said different before the strike, but I definitely know now that it's undemocratic. It's money as buys the lot.'

The strike had brought to prominence a number of men not wise in the ways of the world. More able perhaps than many of their fellows but probably not notably different in their outlook. Men who looked at issues in uncomplicated terms of justice, fair play, and democracy; innocent men in the sense that they had not learned the ultimate cynicism: 'politics is the art of the possible'.

'. . . A BLOODY EDUCATION!'

'Politics' for them was not a question of practicability in a world of conflicting entrenched interests. 'Politics' was a question of what was *right* and what was *wrong*. Issues were judged in terms of morality rather than in terms of expediency. The strike did not change this *outlook*, but it did change their perceptions of the world. The world, they discovered, was not a very pleasant place.

By the end of the strike the distinction between fatalists, reformers, and militants made a good deal less sense than it did at the beginning. But if by the end everyone was *looking* at the world in the same jaundiced way as the militant, they were not all *adapting* in the same way as the militant.

While the militants remained resilient and stoical, some of the fatalists became more than ever convinced of the inevitability of their kind always being at the bottom of the pile – others joined the militants. The reformers went three ways – some stayed reformers, some became militants, and others became fatalists. Events after the strike, when the breakaway union, the Glass and General Workers Union, was flattened between the anvil of the GMWU and the hammer and tongs of Pilkingtons, served only to reinforce these distinctions. Those turned militant found prominent places on black-lists along with the militants. Those who turned fatalist, or remained fatalist, were still working at Pilkingtons along with those who had remained reformers. The story did not end happily for the RFSC or for any of its members. But then their story was not a fairy tale.

Part Three

Chapter Eight

AFTERMATH

The Trades Union Congress – not so magical after all

When the chairman of the RFSC marched the Triplex strikers back into the factory on the morning of 22nd May only the strike was over – nothing else was settled. That afternoon the RFSC travelled to London for a meeting at the TUC with officials of the GMWU.

The meeting that followed lasted for seven hours with a break only for sandwiches and cans of beer. It was not an amicable meeting although according to RFSC members Vic Feather played the role of impartial chairman admirably. The first two hours of the meeting were more or less taken up with a debate over the reinstatement of the Pontypool men – the issue of 'no victimisation' was speedily settled. It was agreed that the GMWU, through its Pontypool branch, would try and get the twenty-eight men reinstated. (The GMWU was not subsequently successful.) The remaining five hours, apart from accusations and counter-accusations, were occupied with a discussion over three proposals put forward by the RFSC.

The first proposal was that the RFSC would be prepared to accept GMWU negotiators if some of their men were co-opted on to the negotiating committee with Pilkingtons – this was rejected by the GMWU. The second proposal, on which the GMWU hedged, was that agreements with Pilkingtons should be put to the shop-floor for their approval before final ratification. The third proposal was that the union side of the Joint Industrial Council should resign and that there should be new elections: this too was rejected. At the end of the talks, the GMWU, in spite of the fact that it had made no concessions to the RFSC, continued to appear conciliatory: it wanted to adjourn the meeting. The RFSC, however, saw no point in continuing and decided to break off talks for the time being – Feather having told the RFSC that he would continue to mediate.

Just over a week later the RFSC contacted Feather again and asked him to arrange a further meeting between themselves and the GMWU. Feather's eventual reply was that the GMWU were willing provided that the RFSC stopped its attacks on the GMWU. The RFSC agreed, and a meeting was arranged for 2nd June – but it never took place. 'A TUC spokesman said that fifteen minutes before Mr Feather was due to meet both sides, Lord Cooper, GMWU general secretary, telephoned to say the union team would not attend. Loyal Pilkington shop stewards with him refused to meet the rank-and-file leaders. Their reason was that these had not carried out their undertaking given last week to stop attacking the union. Since the strike was over, Lord Cooper added, the unofficial strike committee should disband' (*Daily Telegraph*, 3rd June, 1970).

If Vic Feather was not altogether happy about this last-minute rupture: 'I regret the joint meeting did not take place. If the union and the rank-and-file committee had got together tonight they could have reconciled some of their differences,' the RFSC was furious for it believed that it had kept to its undertaking to stop attacking the union. In the event it is likely that Feather was over-optimistic about the chances of reconciliation between the GMWU and the RFSC. The RFSC's price for co-operation within the GMWU was one that the union would not have been prepared to pay, for the price was root and branch reform of the GMWU. Prominent amongst their proposals was that all officials should be elected, periodically re-elected, and subject to recall.

Clearly the RFSC thought it was in some position of strength in St Helens to put forward such demands. The GMWU loyal shop stewards at Pilkingtons however did not think the RFSC was in such a strong position and advised Lord Cooper accordingly: their real reason for not wishing to meet the RFSC for a return match was their view that RFSC support had dissipated after the return to work, and that they were once more in charge of the situation. They saw no reason for negotiating – negotiations implied making concessions, and if they were once more in control of the situation there was no need to make concessions.

By this time it had become very clear to the RFSC that all the TUC intervention had achieved was an end to the strike. By way of issues it had resolved nothing, indeed if anything it had driven the RFSC and the GMWU further apart and diverted the battle to the shop floor, a place where Pilkingtons could hardly be passive by-standers as they had tended to be during the strike. Once the strike was over, Pilkingtons, no matter how hard they might have tried, could not remain aloof from a war which constituted a constant threat to a return to 'business as usual'.

Pilkingtons and the GMWU

In the second week after the return to work Pilkingtons announced 240 redundancies at the Triplex plant due to loss of orders of safety glass from the motor industry. Within a matter of days however the number had been reduced to ninety, and those ninety would be accounted for by natural wastage and by redeployment to jobs in other factories in St Helens. Pilkingtons attributed this change of heart to pressure from the GMWU. When the redundancies were first announced Pilkingtons said that the 'last in first out' principle would be waived so that people who had returned to work during the strike would not be considered for redundancy. The GMWU did not appear to demur – and this gave the RFSC the opportunity to allege that the GMWU was not fulfilling its pledge of no victimisation given at the first TUC meeting. The RFSC thought that Pilkingtons, by discriminating between strikers and strike-breakers when it came to redundancy, were effectively victimising those people who had stuck it out to the end. In the event it did not matter because redundancies were dropped, but while the threat of redundancy was still in the air a number of people took the proffered opportunity to

AFTERMATH

make themselves voluntarily redundant. Included in this number were four members of the RFSC, all of whom had been shop stewards.

As soon as the strike was over Pilkingtons were anxious to press on with changes in wages and negotiating structures, generally to mend fences, and to re-establish 'good relations' with their employees. During the first week back at work it was announced by Pilkingtons and the GMWU that works committees would be set up to discuss wages structures and negotiating procedures. The composition of these committees was immediately a subject of dispute in the Cowley Hill works on 28th May: RFSC members and supporters insisted that nominations for membership should come from the shop-floor and be voted upon by the shop-floor. On 29th May management announced that only accredited union stewards would be acceptable although it had been agreed by loyal GMWU stewards two days earlier that shop-floor workers could sit on the committee.

Although we have no precise information as to what has gone on in these committees, it requires no special genius of insight to suppose that Pilkingtons will have been wanting radically to change and simplify the existing various systems of wage payment by the introduction of 'measured day work' (a system of grading workers according to levels of skill, responsibility, danger, discomfort, etc., and paying accordingly), and to decentralise negotiations somewhat by allowing more bargaining on a plant basis instead of a company basis. Neither does it require any special insight to suppose that changes in the payment system are likely to be resisted by that group of workers currently working a piece-work system, and very well-paid for it. They will be suspecting that any eradication of pay inequalities is likely to be at their expense – that there will be a levelling down rather than levelling up.

While this was going on at Pilkingtons, the GMWU had been conducting an inquiry into the running of 91 branch, the results of which have not been made publicly available. Although the inquiry probably caused some embarrassment in the GMWU because its full-time secretary was elected national chairman of the GMWU during the last week of the strike, its findings would not have been particularly relevant to GMWU members at Pilkingtons, because the GMWU has decided to dismantle the branch and create new ones based on each of the St Helens plants. This looked to be an innocuous reform but events may prove it otherwise, for a plant based union branch coupled with plant bargaining may well make it much more difficult for the national administrative apparatus of the GMWU to control. On the other hand such a change is likely to redound to Pilkington's advantage because on the assumption that each branch is likely to develop different interests associated with the varying technologies of each plant, the chances of all plants coming out on strike together again would seem more remote.

One fascinating question for the speculator must be that concerning the future relationships between Pilkingtons and the GMWU. If, as seems likely, Pilkingtons will try to take a much tougher line in industrial relations, the success of that policy will be dependent upon a toughening attitude on the part of the GMWU towards the firm. Quite apart from the fact that this is just not the GMWU style anyway, the union would find this very difficult to achieve in the short run. The GMWU's problem is that it is very deeply obligated to Pilkingtons: not only did Pilkingtons do a great deal to help out the GMWU during the strike, but after the strike, too, it was Pilkingtons who crushed the Glass and General Workers Union and not the GMWU. It would not be stretching a point too far therefore to suggest that the GMWU owes its continued existence in St Helens to Pilkingtons. If this is near enough the case, how then can the GMWU start getting tough with Pilkingtons?

If the GMWU cannot resolve *this* issue then more trouble at Pilkingtons would not seem to be too far distant, for rank and file GMWU members are unlikely to give the GMWU a second chance. Two things however may obstruct the possibility of the GMWU becoming impotent.

The first is that the GMWU shop stewards may become more militant and force their officials into a similar stance (the shop stewards are unlikely to have the same feelings of obligation towards Pilkingtons as their officials); the second is that Pilkingtons may continue their bolstering act and give the impression that they are under pressure from the GMWU. Either way, those interested in industrial adventure stories would do well to continue looking for the name 'Pilkington' in their daily papers.

The Court of Inquiry

The announcement of the Court came some ten days before the end of the strike, and was accompanied by the customary plea for a return to work. The Court actually started to take evidence in the last week of the strike. The sitting took up two whole days, with witnesses being questioned on occasion by the three assessors and sometimes challenged from interested parties sitting in the body of the court.

Pilkingtons were mainly represented by two of their directors including a member of the family, and the GMWU by a national officer, a research officer, and several local officials. The RFSC briefed a solicitor to appear on their behalf and several members gave evidence. Two members of the RFSC had been instructed by the committee to attend the full hearings and to challenge the other parties where necessary. Pilkingtons and the GMWU submitted both written and oral evidence, the RFSC oral evidence only.

The assessors were Professor Wood, a professor of law at Sheffield University, Mr M. J. Clarke, a personnel executive with the British Steel Corporation, and Mr J. Gormley, a national officer of the National Union of Mineworkers. During the court proceedings they obviously tried very hard to be reasonable and tolerant with witnesses, although a certain edge appeared in Mr Gormley's voice as he questioned the chairman of the RFSC. Overall, the tone of the hearings was amicable and conciliatory: it seemed very

THE COURT OF INQUIRY

unreal considering the height of tension prevailing in St Helens, an unreality which was a product of the *idea* of a court. All three parties were impressed by the judicial nature of the hearings and saw the assessors as independent people who could be relied upon to come to a reasoned judgement. The witnesses were therefore on their best behaviour and anxious to impress the Court as reasonable men. The result was that they minimised the differences between them – Professor Wood could hardly be blamed for saying in his summing-up: 'We have heard all three sides say in many respects exactly the same thing, and when one examines the area of difference, it is very small indeed.'

In the circumstances the eventual report of the Court said nothing of note, nothing that could not have been culled from a judicious selection of clippings from *The Financial Times* or *The Guardian*. The court, predictably, found fault in all three parties, with the weakest party – the RFSC – coming in for the strongest criticism. The RFSC was soundly admonished for its failure to accept the verdict of the 'parsons' poll' and for its 'lack of mature judgement and sound leadership . . .' in respect of that failure.

Pilkingtons were found to be somewhat paternalistic in their approach to industrial relations, insufficiently aware of communications problems, and too leisurely in their attempts to iron out anomalies in the wages and negotiating structures. The GMWU was let off very lightly: it was urged to re-organise its St Helens branch on plant bases, to encourage participation in its decision-making processes at branch level, and to be 'less ready to accept the Company dictated pace of reform and change whenever it feels its members' interests are at stake'.

It is tempting to suggest that the report, published some two and a half weeks after the end of the strike, was redundant. It recommended nothing that Pilkingtons and the GMWU had not already considered, its apportionment of praise and blame was irrelevant in that it applied to a situation already receding into myth and folk-legend, but most of all it completely failed to come to grips with what had been the realities of that situation. One cannot help feeling that a study of the Court of Inquiry would have been more revealing of Courts of Inquiry, than was the Court of Inquiry's study of the Pilkington strike.

To the extent that Courts of Inquiry are political devices to resolve embarrassing disputes, then this particular court was redundant. The announcement of its establishment a week or so before it began to hear evidence did not secure a return to work, in spite of a plea from the Minister of Employment and Productivity. A repetition of the plea on the first day of the hearings met with a similarly negative response. The fact of the matter was that the only party that could have called off the dispute, the RFSC, hardly noticed the court at all. Its pleas were never discussed in committee, and it was only at the last moment that it even decided to give evidence. It was not that there was any disrespect intended – it was just that the Court did not seem to matter because it seemed so remote from what was going on in St Helens.

But if the *Court* was redundant, its *report* was probably not. To the extent that Pilkingtons and the GMWU took it seriously, it will have proved important. Partly because it reiterated much of what the firm and union had said in evidence – reiteration by an apparently independent and 'truth-seeking' body may have reinforced their views of the strike and its causes; and partly because its recommendations may have been taken to heart. As far as the RFSC was concerned however, the report *was* redundant: after two abortive TUC meetings the RFSC was in no mood to listen to advice that urged them to stay in the union and work for change from within. If Courts of Inquiry effectively amount to conciliation procedures, and there can be few doubts in anyone's mind that that is what they are intended to be, then this one was unsuccessful. There was no question of a need for a reconciliation between Pilkingtons and the GMWU for there had never been a rupture, and by the time the Court's report appeared events had overtaken the possibility of a reconciliation between the GMWU and the

RFSC. With the men back at work the GMWU apparently saw no reason for making any conciliatory gestures toward the RFSC.

The crux of the matter as far as this particular Court was concerned, and indeed for Courts of Inquiry generally, is that conciliation seems to necessarily imply a basic unity of purpose between the disputants. Certainly the tone of the report into the Pilkington strike strongly inferred that, given an increased dosage of goodwill all-round, all would be well in the future. As we shall see in the concluding chapter this is an ill-found assumption. The Court of Inquiry never came to a full understanding of what was happening in St Helens: if it had it would have found little evidence of basically harmonious relations between Pilkingtons, the GMWU, the RFSC, or the rank and file striker.[1]

The Glass and General Workers Union

On Friday, 5th June, the RFSC handed into Pilkingtons just short of 3,500 forms instructing Pilkingtons to discontinue deducting GMWU dues from wages. Shortly thereafter the RFSC reconstituted itself as the Pilkington Provisional Trade Union Committee and started to have weekly meetings of its supporters. Towards the end of the following week Pilkington personnel officers were busy checking the signatures of these forms with the signatories. GMWU shop stewards were also active, in company time, trying to persuade people to re-sign with the GMWU.

At this stage the new committee were still hoping eventually to join another union – the Transport and General Workers Union. They were sustained in this hope by a full-time official of the T&GWU who told them at a packed meeting that a coach-and-horses had been driven through

---

1. We have relied, in this section, on the *Report of a Court of Inquiry under Professor John C. Wood into a Dispute between Pilkington Brothers Ltd., and certain of their employees*, London, HMSO, 1970, and on notes taken at the hearings by our colleagues, Gary Littlejohn, and Gideon Ben-Tovim.

the Bridlington Agreement (a TUC agreement governing relations between unions with respect to the recruitment of each others' members), and that he was confident that they would be able to join the T&GWU. This advice, as the new committee was later to discover, was both bad and ill informed. The fact of the matter is that a coach-and-horses has never been driven through the Bridlington Agreement. The TUC disputes committee, which administers 'Bridlington', has not once since 1945 made an award in favour of a union recruiting the lapsed members of another. (It is almost impossible to resign from a union – the Pilkington 'resigners' would have been regarded by the TUC disputes committee as lapsed GWMU members.) There was, in short, never much of a chance that the T&GWU would have taken the rebels in.

By the end of June the Pilkington Provisional Trade Union Committee realised that no other large union would accept them into membership and decided to form their own union – the Glass and General Workers (GGWU). Membership cards were printed and dues collectors appointed. Branch meetings were held weekly with attendances varying between 150 to 200 men, with the GGWU claiming a membership of 5,000. Very large numbers of the new union were attending the weekly meetings, for since the meetings were held at the same time each week, and since shift workers would be on different shifts each week, the same 150 to 200 men could not possibly attend every meeting. The new union started off with some considerable enthusiasm. The actual conduct of the meetings was testimony to this – a microphone was placed in the centre of the floor, and 75% of the duration of the meetings was taken up with contributions from the floor. The meetings invariably ran over time.

The new union was not, of course, recognised by Pilkingtons, although informally there had been recognition to the extent that in several plants managements had agreed to discuss issues with GGWU representatives (on the understanding, naturally, that such discussions did not amount to

GLASS AND GENERAL WORKERS UNION

recognition.) With regard to recognition the GGWU's policy was that it did not want to be rushed into a recognition dispute. Its strategy was to consolidate and to wait for a really 'good' issue before pressing its claim. Some of the members however were impatient and on 5th July forty men walked out at the Triplex plant after management refused to meet their representative over a bonus dispute. Pilkingtons promptly threatened to sack the men unless they returned to work, and the GGWU found itself obliged to advise their members to return.

During July things seemed to be running well for the GGWU. Early in the month they had been approached by lay executive members of the 7,000 strong, TUC affiliated union, The National Union of Domestic Appliance and General Metal Workers (NUDAGMW) with a view to a merger. Approaches had also been made with a view to membership from other plants in other firms organised by the GMWU both in St Helens and outside. At this time GGWU looked well placed to engage in raids on many other GMWU-organised firms all over SW Lancashire and not a few in other parts of the country. In all the GGWU received well over 100 inquiries, most of them from GMWU members. At that time it looked as though the GGWU could grow to a membership of 25,000 members very quickly.

By the end of July the GGWU decided to start its campaign for recognition from Pilkingtons. On Monday, 27th July, a leaflet was issued to its members stating that a work-to-rule would start at Cowley Hill one day after the national dock strike was over – the work-to-rule duly started on 4th August. By that time a further GGWU meeting on the day previous had decided to add on an overtime ban as well.

On the day that the work-to-rule and overtime ban started – Tuesday 4th August – a GGWU member was suspended at Cowley Hill works for an infringement of works rules. An official of the GGWU tried to contact Pilkingtons' head office industrial relations staff to discuss the suspension but they refused to talk. On Wednesday GGWU officials tried to represent the man when he appeared before the works manager, again without success. At a mass meeting of members from the Cowley Hill works on Thursday afternoon a resolution was put from the floor to the effect that there should be a three-day token strike in protest against Pilkingtons' refusal to allow a GGWU representative to appear with the suspended man.

Pilkingtons' response was immediate. On Friday, 7th August, the firm said that if there was not an immediate return to work all those on strike would be considered to have terminated their employment. By the Tuesday of the next week Pilkingtons had dismissed some 480 people but were offering re-employment to those who had been sacked. Attempts by the GGWU to spread the strike to other plants on Monday, 10th August, and on Tuesday, the 11th, met with very little response, although some GGWU executive committee members felt morally obliged to support the strike. By Wednesday, 12th August, practically all of the remaining members of the RFSC had been sacked.

The GGWU now demanded reinstatement for those who had been dismissed (re-employment meant that former employees would be treated as if they were new employees and would forfeit pension rights and redundancy compensation if they were to be made redundant) but had very little with which to back up such demands. On Friday, 14th August, the GGWU had obtained the support of the Liverpool dockers who started to 'black' Pilkington glass, but by Tuesday, 18th August, the dockers had lifted their ban after they had been informed by a Pilkington director that the Pilkington workers that had been sacked would be reinstated. This was not the case, Pilkingtons had not changed their tactics – the director had made a mistake: he meant re-employment. Pilkingtons made it clear in the previous week that re-employment would be selective, i.e not everyone would be offered re-employment, and that a condition of re-employment would be membership of the GMWU.

GLASS AND GENERAL WORKERS UNION

On Friday, 14th August, the GGWU tried an additional tactic – a writ was served on Pilkingtons by the GGWU solicitor suing them for wrongful dismissal and for damages of £750. The GGWU solicitor was confident that this would bring Pilkingtons to heel, that the firm would not want the publicity attendant upon a court case, and that everyone would therefore be reinstated within a fortnight.

A fortnight passed and Pilkingtons showed no signs of being frightened by the prospects of a court case. By the beginning of September there were 250 men as sacked then as they had been three weeks earlier.

At the time of writing – early September – the Glass and General Workers Union remains in abeyance, its assets frozen by the executive committee. The men who have been sacked continue to regard themselves as being out on strike, but once more are marching under the banner of the RFSC. If there are prospects of a large proportion of that 250 being re-employed at some time in Pilkingtons, there are a number, perhaps as many as 100, who stand very little chance, and a smaller number of perhaps a dozen who stand no chance. Pilkingtons have successfully 'nobbled' that group of people who will probably have been seen as standing between themselves and a brighter future.

The re-born RFSC has not as yet given up the struggle and continues to be an irritant to Pilkingtons and the GMWU. Men still picket the Cowley Hill gates from time to time, and delegations travel the country trying to whip up support. Workers in odd parts of the country have been persuaded to 'black' glass, and resolutions continue to be passed by shop stewards committees. Strikes involving GMWU workers are a natural target for the new RFSC, and delegations are duly sent to warn these strikers of the 'treachery' of the GMWU. It seems likely that by the time this manuscript has been converted into a book the Glass and General Workers Union will have been wound up. It also seems likely that there will be a large number of workers in Pilkingtons, ex-members of the GGWU, who will not have rejoined the GMWU. But most likely of all is that Gerry Caughey, chairman of the RFSC and GGWU, and John Potter, treasurer of the RFSC and the GGWU, will be on the dole.

The problems of the breakaway union

If the members of the RFSC had read Shirley Lerner's *Breakaway Unions and the Small Trade Union*[2] they might possibly have had second thoughts about setting up the GGWU. On the other hand it may well have proved that the political realities of their situation left them with little choice if they were to remain a credible source of leadership. After all they had said about the GMWU and the support their remarks appeared to have, they would have found it difficult to have stayed in the GMWU and carried their supporters with them. But these considerations apart, there were a large number of RFSC members who so detested the GMWU and all it stood for that it was *personally* impossible for them to compromise. All the 'rational' calculations in the world counted for nothing after the GMWU refused to meet them for a second time at the TUC.

When the RFSC decided to set up the GGWU they were not aware of all the problems they would face. They knew that they were not going to have an easy time, they knew that they had formidable enemies and no powerful friends, but they did not appreciate the near certainty of their destruction. If they had had an education during the strike it had only been at the primary level. If they have now graduated their degrees are worthless in terms of future application.

Since the end of the First World War there have been a number of secession movements within different trade unions but *none* of them have been successful. At least one, the Union of Railway Signalmen – a breakaway from the National Union of Railwaymen in 1924 – continues in existence but has never gained any negotiating rights. The

2. George Allen and Unwin, London, 1961.

THE PROBLEMS OF THE BREAKAWAY UNION

National Amalgamated Stevedores and Dockers Union, a breakaway from the T&GWU in 1923, also continues: it too has never secured national negotiating rights.

The established trade union movement has naturally never approved of breakaways, and the fact of this disapproval has always proved sufficient to squash the rebels. The established unions' tactics have either been to bring pressure to bear on employers' associations or to refuse to co-operate in negotiations if the firm recognised the breakaway. The GMWU might well have adopted this position at Pilkingtons.

The GMWU, just prior to the strike, had roughly 11,300 members at Pilkington plants throughout Britain, of whom some 7,400 worked in St Helens. These workers were represented nationally by the GMWU on the Joint Industrial Council. Suppose that the GGWU had managed to get a clear majority in St Helens of, say, 5,000 workers, that would still have left them with a minority at a *national* level and perhaps enabled the GMWU to squeeze them out. As it happened the new union did not have a majority in St Helens and did not therefore have much of a chance unless the GMWU had made a crass blunder and thus helped the GGWU to sweep the board. The GGWU of course was waiting for just such an event, indeed were fully expecting it to come out of the wage restructuring. This of course was to reckon without Pilkingtons who were also alive to such a possibility and perhaps prepared to save the GMWU from yet further ignominy. As Shirley Lerner put it: 'Employers' associations oppose breakaway unions and new unions almost as strongly as do the established unions. Recognition of a "wrong union" could cause a greater amount of industrial strife and be more costly than non-recognition' (p. 195). This almost certainly summed up Pilkingtons' position and explains why they acted so promptly to crush the GGWU at the first opportunity.

Breakaway unions, if they have to contend with established powerful groups, also have to contend with their actual and potential membership. They have to contend with the fact that there is a difference between being a trade union member and being a trade unionist. They have to allow for the fact that life is not a continuing crisis, and that it is crises which turn trade union members into trade unionists, that it is crises which transform seemingly passive clients into militant activists. Leaders of breakaways are no doubt subconsciously aware of this and accordingly attempt to maintain crisis pitch. But this cannot be kept up for long and attempts to achieve it are likely to lead to the isolation of the activists. And once they are isolated they are easily put down – as were so many of the Fords shop stewards at Dagenham in 1962 when the firm and the unions together got rid of seventeen militants.

Seasoned militants are not likely to be the sort of people who lead breakaways for the process of seasoning means amongst other things acquiring a sense of political realities which transcends feelings of moral outrage. The militant is aware of how difficult it is for the union hierarchy to control his actions and that a good shop stewards committee or rank and file organisation is better equipped to deal with life on the factory floor than the brave but doomed heroism of the breakaway union.

Chapter Nine

CONCLUSIONS

No two strikes are identical. What happened in St Helens in the Spring of 1970 will never happen again, either there or anywhere else. But there are elements common to many strikes. The Pilkington strike was a wildcat – it was spontaneous and completely unpremeditated – and developed into a protracted struggle. These features, either taken separately or together, are far from unusual.

In this chapter we will attempt to generalise from our own study of one particular strike to other strikes having similar characteristics. While we cannot be certain that the conclusions we draw from the Pilkington strike will stand up to the sort of generalisations about to be emptied on to the reader, we nevertheless feel bound to suggest that our bucket has dredged up something of more than local importance. Just how important is for our readers to judge, and for other social scientists to confirm or dispute by further research.

Strikes are 'normal'

Very few people at Pilkingtons had been expecting a strike. How and why the strike spread from a small beginning remained a mystery to most of the participants. Shop stewards, rank and file workers, and managers all confessed to an inability to understand the events of the first weekend. The dispute over a clerical error which sparked it all off surprised nobody, for small localised conflicts were not unknown in the Flat Drawn department or in other departments in other plants. What *was* so amazing was the way it became converted into a demand for a large increase on the basic pay and the way it snowballed right around St Helens.

In what was an otherwise extremely complex series of

CONCLUSIONS

events only one thing stands out with clarity: there was no organised plot. The strike had not been engineered by a group of subversives who had deliberately infiltrated Pilkingtons in order to undermine the economy. Cloaks, daggers, Kremlin agents, the church of Rome, and small groups of 'politically motivated men' – all those mythical progenitors of natural and political 'calamities' belong firmly in the pages of Ian Fleming and Dennis Wheatley. Drama there was in St Helens – but it was unscripted. Such script as there was, was made up by people as they went along: the strike in its beginnings was a genuine spontaneous movement. In the early stages many workers did not know why they were striking – it was during the process of spreading through the factories that the strike acquired definite objectives.

Previous explanations of this type of wildcat have treated it as an emotional explosion following upon pent-up grievances that have not been resolved. Yet while this theory fits approximately the department in which the strike started, the same could not be said for all of the departments in all of the works. Morale in the Pilkington factories did not appear to be unusually low in the period immediately preceding the strike. If top management had been warned that they were sitting on a 'powder keg' and had noticed that they were '... getting rather less cooperation in change than (they) had in the past', they certainly weren't prepared for what they woke up to on the morning of Saturday, 4th April. There had been no sudden acceleration in labour turnover, and the incidence of localised disputes had shown no sharp upswing.

Most of the workers had not anticipated the strike either: they had seen no build-up of tensions or problems that had led them to want to strike or to believe that one was imminent. Problems and grievances certainly existed. Dissatisfaction with wage levels was widespread. In some departments there had been almost continuous friction between labour and management. Some employees (including most of those who became members of the RFSC) had felt that quite a lot of things were badly wrong at Pilkingtons. But, if it could not be claimed that the firm was purring along in a state of glorious harmony, there was certainly nothing to indicate that on the day the strike started, grievances had been more numerous or more deeply felt than they had been in the preceeding months.

This means of course that workers can be drawn into a strike without being conscious of an exceptionally wide range of grievances, and without being subject to unusual stress on the shop-floor. A strike, in other words, can gather momentum under 'normal' working conditions. The prospect of a sizeable pay increase, apparent support from workmates, and advice from shop stewards, can together encourage people to join a strike and thereby add to its momentum – all without any change in the feelings and attitudes characteristic of previous months. The actual dynamics of the act of striking have already been discussed in Chapter Three. At this stage we would merely remind readers that those people who actually set a strike in motion undergo rather different experiences from those who subsequently join it, and that a large-scale strike can be triggered off by the decisive action of only a small number of men. This however will not always be the case. It has to be remembered that the great majority of the Pilkington workers had never before been involved in a strike-prone work situation. In firms where workers are more strike-experienced it is much less likely that a departmental-based dispute would have a bush-fire effect – even if it were to be converted into an issue of wider application.

In contemporary Britain strikes have been defined as a 'problem', and the attachment of this label has created the allied impression that strikes are aberrations and atypical events which must therefore occur only amongst aberrant workers or under atypical conditions. It has accordingly been assumed that if certain features of the industrial landscape were remodelled the 'problem' could be buried.

This conception of strikes may well be false. The implication of our observation – that the Pilkington strike occurred

CONCLUSIONS

in normal conditions – is that strikes generally should be regarded as *normal* features of industrial life in those sectors of the economy where large numbers of workers are gathered together under one roof.

The Pilkington workers' complaints about their jobs were mainly centred around money: 'We have to work a lot of overtime to get a decent living wage,' and of course during the strike a return to work was made conditional upon the payment of cash. This suggests that work was primarily regarded as a matter of exchanging time and labour for cash – not as a source of interest or stimulation, or a moral obligation to the employer. This attitude to work is far from uncommon amongst semi- and unskilled manual workers in large firms. Employers treat labour as a commodity and workers regard their labour-power in a similar way.

This is not to suggest that people look only for financial rewards from their work; there is in fact more than enough evidence to show that ideally workers would like to derive a much wider range of satisfactions. The suggestion is simply that as a matter of fact the nature of work is such that money is about the only reward that can be attained.

This instrumental approach towards work may have several repercussions for industrial relations. Workers with such a disposition may well use whatever tactics appear to be most effective to maximise their earning power. They may bargain through their union officials or shop stewards, they may work-to-rule, go slow, ban overtime, or strike.

From the point of view of explaining the way in which the Pilkington strike gained momentum over the first weekend, the implications of this instrumental approach to work are obvious. If there is no sense of moral obligation towards work, no feeling of personal involvement, then the possibility of a quick financial return will be sufficient to get people outside the factory gates. This certainly seemed to be one of the features in the first crucial forty-eight hours of the Pilkington strike, for, to repeat, there did not seem to have been any accumulation of tensions throughout all the plants in St Helens.

Some sections of industry are more strike-prone than others. The breadth and depth of bonds that tie individuals to their jobs and their employers will vary, and so the extent to which apparent trivia such as the miscalculation of bonuses are likely to spark off something larger will also vary. There can be no doubt that some industrial conditions are more conducive to strikes than others, but to claim that strikes can be normal is not to deny this. The fact remains that in much of large-scale industry millions of workers are employed under conditions in which periodic upsets are bound to occur, and when they do there is the continuous possibility of a wildcat strike.

Pilkington management admitted that problems did exist: clerical errors in the calculation of wages were far from unusual; anomalies did exist in the wages structure; some jobs were not particularly pleasant, and others were either distinctly unpleasant or dangerous. But management was also quick to point out that most other large firms had identical problems, and anyway what could be done about them? The answer is: very little. And this is precisely why strikes can be perfectly normal events, conducted by perfectly normal people, in perfectly normal circumstances.

An industrial protest movement

In the preceding section, by concentrating on the idea that many manual workers have a pecuniary attitude to work, we have glossed over several important issues. As we have pointed out earlier a worker sells more than his labour when he undertakes to work for an employer – as well as undertaking to perform certain tasks he also undertakes to abide by a set of rules, he submits to a system of authority. From this it follows that whether the worker likes it or not, whether he is primarily interested in money or not, he nevertheless sacrifices certain areas of freedom, a sacrifice furthermore that is not likely to be lost on him. It is quite unrealistic to suppose that because a worker works only for money he accordingly shuts off his mind to his daily experiences at the

CONCLUSIONS

factory. If he treats his *labour* as a commodity it does not follow that he expects himself, as a person, to be treated as a commodity. Neither does it follow that he will be prepared to put up with anything if the money is right – if car workers are the best paid workers in Britain they are also the most militant.

This question of authority has not been lost on employers. They have been well aware that how authority is exercised may make a big difference as to whether workers co-operate willingly or grudgingly. Many large firms in Britain and elsewhere spend large sums of money every year in an attempt to secure loyalty to the firm: to what effect is another question. This is not to suggest incidentally that these attempts to obtain willing consent to the established system of authority are operated cynically. If some are, there are others that patently are not. In the case of Pilkingtons there can be little doubt that the family has been sincere in its belief that it is not *just* in business to make money. There can be no doubt either of the sincerity of their belief that the *whole* man has to be catered for, that their interest in their workers *does* go beyond the appropriation of their labour.

Yet sincerely held or not, it is indisputably true that manifestations of concern have the effect, intended or not, of maintaining the existing system of authority. Welfare schemes, fringe benefits, company magazines, etc., all help to sugar the pill, to obtain acquiescence in the status quo. But 'brand loyalty' as many an ad-man and football team has discovered can be very fickle: there can be no guarantee of continuing allegiance – assuming it to have been attained in the first place.

It must of course be highly questionable in an age of large-scale organisations whether attempts to get company loyalty will meet with success. Loyalty is very much a *personal thing*, and if 8,000-odd people are concerned the 'laying on of hands' becomes an impossible job. Attempts to achieve it by the odd walk around the shop-floor are more likely to be received with ribaldry than with deferential affection precisely because they are *odd*, because they are unusual, because they have been preceded by all the tidying and cleaning operations that go on before the boss appears. 'A *real* boss,' it might be said, 'would want to see us as we really are.'

At Pilkingtons the old order of autocratic paternalism had broken down, but its passing, as we have seen from RFSC members' comments, resulted in no little ambivalence. Amongst some of them was the implicit belief that if the firm had lived up to its traditions of being a family concern, that if it had shown more interest in its employees as persons, there would never have been any trouble. There was a belief that the family had been misled by its lower echelon managers and supervisors. Coupled with this however was a marked hostility to the way Pilkingtons were seen as dominating St Helens. If there was an affirmation of a traditional attitude to authority in which there was an acceptance of their 'station in life', there was a contradictory affirmation of the modern attitude to authority, an attitude of distrust, some resentment, and a demand that it be accountable to those who are its subjects. The strike, clearly, will have tipped the balance towards the modern, towards a recognition of a conflict of interests rather than a harmony of interests, but not without some misgivings on the part of some Pilkington workers, particularly those older workers who will have spent a large proportion of their working lives with Pilkingtons.

The modern attitude to authority was well expressed by those workers who, in the act of striking, experienced feelings of elation and liberation. These feelings are, we suspect, extremely common amongst those who spark off a wildcat strike. They are feelings characteristic of those who have momentarily dissolved the shackles of authority. But if these feelings are positive in the situation of striking, they are negative in the longer run for while they are saying 'down with this', they are not at the same time saying 'up with that'. No alternative is being offered, so that when the strike is over they return to the same fundamental system of authority that existed before the strike. At *Pilkingtons* this is less likely to be the case. If they do not shed overnight their patriarchal attitudes towards their employees, the fact that the company is about to become public and the fact that career managers will continue to have more sway over company policies, will almost inevitably mean that attempts to shore-up the old order will come to grief. A reliance on a 'store of goodwill' still remaining amongst some employees is also unlikely to forestall this change, for the younger generations of workers will, in tune with the climate of the era, be predominantly instrumental in their attitudes towards work. In this sense the strike at Pilkingtons marked the end of one era and the beginning of another: at Pilkingtons there will be no return to the 'good old days'.

How was it that the old order at Pilkingtons crumbled? This is not an easy question and there is no simple answer. There are unquestionably a number of reasons, but there is no way of deciding which were the most important. One is certainly that which we have talked about at some length in the previous section – the changing attitude to work which is probably related to the way in which monetary values increasingly pervade so many aspects of social life in advanced capitalist societies. Another reason may be associated with the 'strike-conscious culture' which will be discussed in the next section. Perhaps the main reason has to do with the erosion of the insularity of St Helens and the growth of manufacturing industry in areas easily accessible from St Helens. Pilkington dominance of the St Helens labour market is no longer of the same importance – striking is not the dangerous business it once was, or appeared to be.

In an earlier chapter we suggested a similarity between the Pilkington strike and a peasants' revolt. While we do not intend this term in an insulting way, the fact remains that there was a close resemblance between the attitudes of some of the RFSC and those of rebellious peasants in pre-industrial societies. A common feature of many peasant revolts was an appeal to established authority to curb the excesses of underlings who had visited injustices upon loyal

AN INDUSTRIAL PROTEST MOVEMENT?

subjects. The significant feature of such revolts was the appeal to established authority – authority as such was not being challenged.

When the RFSC first formed it saw itself primarily as a 'ginger group' within the union – it aimed to draw the attention of the GMWU hierarchy to the views of the strikers in the hope that sympathetic action would be taken. At no stage was it the RFSC's aim to attack the trade union movement and replace it with a different type of labour organisation. Indeed, after it had despaired of the GMWU, the RFSC's response was to transform itself into another trade union with conventional trade union aims and policies.

With regard to Pilkingtons we have already pointed out that a number of RSFC members were not only willing to differentiate between the family and the organisation, and hold the underlings largely responsible for the strike, but were also protesting that the family did not see enough of its employees: the unspoken caveat being that if they did things would be better.

The whole range of attitudes of RFSC members were redolent with protest, and many of their appeals were significantly addressed to established authority, e.g. the appeals to the mayor of St Helens and to the Minister of Employment and Productivity. There was a strong belief that if only people could be got to listen the 'powers that be' would recognise the justice of their claims and the dispute would be satisfactorily resolved. We have no way of knowing the extent to which this sort of protest was unique to Pilkingtons, but insofar as a large number of people every year find themselves leading an unofficial strike for the first time in their lives (on the assumption that every year a firm finds itself featuring in the strike statistics for the first time), then these sorts of attitudes must be quite common.

Ours is an individualistic era in the sense that great value is placed upon the development of personal qualities. Yet many people, perhaps even a majority, find themselves in jobs which leave almost no scope whatever for individual initiative. And not only that, they find themselves doing jobs

CONCLUSIONS

in an environment which reduces them as persons to cogs in a wheel, to numbers on a clock card. The arrangements of technology and authority require unthinking obedience. Little wonder then that wildcat strikers sometimes talk as if they have 'done something big for the first time in their lives'. Such people are proclaiming their humanity and protesting that their work situation denies it.

It seems to make sense therefore to characterise strikes as a form of protest movement, though perhaps the word 'movement' is misplaced since that implies an organisation with definite aims – a wildcat we have seen has no initial organisation and often only acquires concrete aims in the launching process. Frequently, therefore, all that is required to set a strike in motion is an apparently trivial grievance. It is not unlike a row with the wife – a row that starts over something that would ordinarily seem ridiculous, but as it progresses more fundamental issues are raised which put in question the very idea of marriage. Likewise a strike, whether the participants are always aware of it or not, puts in question the whole basis of employment. It is in this way that strikes are protests.

A strike-conscious culture?

We have suggested that the Pilkington dominance of St Helens has been steadily eroded over the last decade, but this does not help to explain why the strike took place in 1970. Why not in 1966, or 1969, or at any other time in the last five years? This is yet another of those questions that cannot be fully answered. There is nevertheless one factor which we think might have been of some importance: a 'strike-conscious' culture.

If St Helens was living on a staple diet of strike for seven weeks in the Spring of 1970, Britain as a whole has become accustomed to consuming strikes with its cornflakes over the last five years or so. Strikes have come to be considered as common and effective means of getting bigger slices of cake.

A STRIKE-CONSCIOUS CULTURE ?

During the Labour Government's period of office between 1964 and 1970 the increasing rate of unofficial strikes was raised to the status of a 'serious economic problem'. Politicians, ably assisted by the mass media, successfully created the impression that if only strikes could be made to go away they would take with them most of Britain's economic ills. Where before this period strikes were mainly of interest to afficionados and industrial relations 'experts', they became a matter of widespread popular and political debate. The view that British industry was especially strike-prone became firmly embedded in popular consciousness.

During 1969 and the early months of 1970 the Government's previous attempts to regulate prices and incomes were first relaxed and then virtually abandoned – giving rise to an apparent wages explosion. Many of the wage increases *appeared* to have been won as a result of actual or threatened industrial action.

In our investigation it became apparent that many Pilkington workers felt that their wages were lagging behind. Although figures were produced which indicated that their earnings were on average in excess of what most other British workers earned, the majority of the strikers had a quite different impression. Frequent references were made to big increases gained by workers in other industries, particularly the £4-odd rise gained by Ford workers just down the road in Liverpool without a strike.

It seems not unlikely that an *apparently* increasing strike rate, particularly if strikes seem to result in significant gains, may have its own snowball effect throughout the economy. Paradoxically, it may be that the more attention is drawn to strikes the more frequent they become. This type of effect is always difficult to substantiate. But as regards strikes, which seem to be increasing in frequency amidst mounting publicity, the theory at least deserves further attention.

But this is to mention only one aspect of the 'strike-conscious' culture. Another aspect, and a much more disturbing one from the point of view of those in the trade union movement, is the apparent support (as indicated by

CONCLUSIONS

opinion polls) of rank and file trade union members for a 'get tough policy' on the movement. *If* the opinion polls are to be relied upon, it would appear that many workers have been convinced that strikes are an 'evil' and that trade unions have become more powerful than the employers. Such views are either inappropriate or plainly mistaken.

There is not much point in moralising about strikes when they are part and parcel of the economic system, and to suggest that trade unions have become more powerful than employers would be laughable if it were not taken so seriously. The day to take such suggestions seriously will be the day when a shop steward has the power to sack his managing director.

What is so often forgotten is that the trade unions are primarily *defensive* organisations: where employers are in a position to set economic change in motion, the trade unions can normally only *react* to change. If it is certainly true that trade unions and workers have become more powerful in a situation of full employment and of domination of markets by a few firms, and if it is also true that in such a situation trade unions and workers can exercise some restraint over employers, it might usefully be remembered that it is *restraint* that is exercised. Where the trade union movement has chosen to work within the established economic system it could hardly be otherwise.

One of the acid tests of power is the distribution of economic resources, and on this count alone it is clear that the trade union movement is no more powerful now than it was in 1900. Since 1900 there has, contrary to popular belief, been no significant redistribution of the national income. If everyone is better off, much the same income gaps remain between various groups of the population. This state of affairs could hardly exist if the trade unions were now more powerful than employers. Obviously some aspects of the strike conscious culture are based on a false view of economic and political realities.

The escalation of conflict

No one at Pilkingtons expected a seven week strike – most people expected it to be settled within a few days or a week at the most. In the event conflict escalated as each week seemed to add a new dimension. Why did the strike develop into a protracted struggle?

One thing that struck us most forcibly was that the public image of the strike as conveyed by television and most of the press – not to mention the proceedings of the Court of Inquiry – was its failure to come to grips with the real nature of the dispute. Popular press reports were misleading not only because they tended to spotlight the sensational, but also because they tended to suggest that there were definite issues over which the parties were dead-locked. Generally speaking the line was that the strike continued because the parties were haggling – at a distance – over the price of a return to work. The same issue was dominant at the Court of Inquiry.

This image was thoroughly misleading, for a fundamental feature of the strike was that there were no arguments centred around a clear set of issues. If there was a dominant view amongst the rank and file it was that the strike should be ended as quickly as possible on the best possible terms. Everyone else recognised wages as being important too, but each group of generals saw other questions as having almost equal importance.

A key issue for the GMWU was the need to protect its established negotiating position in what it considered to be the long-term interests of its members by ensuring that no concessions were made to what it saw as 'power-hungry demagogues', 'political subversives', and the 'anarchic rule of the mob'. To the RFSC it was of great importance that they should be recognised as speaking for the majority of the strikers and that there should be democratic participation in the taking of decisions. Pilkingtons developed a major concern that it should not do anything to permanently jeopardise relations with any party with whom it might have to

CONCLUSIONS

conduct future negotiations.

While each party had its own conception of what were the main issues, each also tended to refuse even to acknowledge the relevance of issues that were considered important by the others. This, essentially was what the prolonged strike was about and why it proved so hard to settle.

The strike never took the form of a straightforward debate, there was never any pitched battle or confrontation – and there never could have been because they were not agreed as to what to fight about. In withdrawing to positions based upon quite different issues the participants were not deliberately attempting to be obstinate and intransigent. What was sincerely felt to be a real issue by one party was genuinely felt to be irrelevant by others.

That the contenders were fighting quite different wars helps to explain the liberal generation of conspiracy theories. The firm and the union both suspected that they were being challenged by a subversive plot in which alien political forces were playing a manipulative role. The RFSC, in turn, strongly suspected that it was facing an unholy alliance of the firm, the union, and (at times) the police in which plots and strategies to deceive the strikers were being hatched. Propaganda issued by both the RFSC and the union sometimes pushed the view that each was facing conspiratorial opponents.

In fact these theories were largely mythical. It would really have been difficult to imagine a more tactically inexperienced and politically unsophisticated group than the RFSC, and the blundering character of the union's tactics made it an unlikely partner in any subtle plot. If Pilkingtons did from time to time work closely with the police they were only exercising their rights as property owners. The theories, if mainly mythical, were not without consequence for they persistently inflamed the feelings of the respective parties.

The polarisation of 'the generals' into different camps based upon different issues, of which the generation of conspiracy theories was symptomatic, was what made the strike so difficult to settle. An explanation of the prolongation of the dispute requires then a prior explanation of why the polarisation initially occurred.

The crisis situation created by the sudden eruption made certain groups more conscious of issues of which they had previously been only dimly aware. The union's officers, who had not before doubted the security of their positions in St Helens, became aware that they were being threatened: they accordingly felt that the situation needed to be 'brought under control'. Amongst some of the strikers, from whom the membership of the RFSC was subsequently drawn, there grew an awareness that there was no way in which they could ensure that they were consulted before decisions were made which affected them: they started to think that the people who were supposed to represent them would 'sell them down the river'.

The union saw, and reacted to, the situation very differently from its members. Union tactics were at the beginning, to put it gently, indelicate. The way in which the local branch positively declared itself to be behind the strike, only to be followed by a public repudiation from one of the GMWU national officers, could not have demonstrated more effectively the disparity between the views of the strikers in St Helens and those of the union's national hierarchy.

The fact that the GMWU is renowned for its addiction to the 'national interest' ideology and for its unequivocal opposition to unofficial strikes almost certainly influenced GMWU tactics, and thus contributed to the escalation of the strike. A union more sophisticated in its strategy and possessed of a different ideology might well have brought the strike to a more rapid conclusion.

But, as we have explained in previous chapters, union officials will almost inevitably react to the outbreak of an unofficial strike with feelings different from those of their members. Wherever there are large and well-established unions which have committed themselves to definite ways of conducting business, that have entered into relations of mutual regard with employers and governments, then con-

CONCLUSIONS

ditions will always exist in which, given a crisis, an awareness of a gulf between trade union officials and rank and file members can easily emerge.

It was the developing awareness of this gulf that set the strike on to an apparently endless escalator. While the strikers thought the main issue was a pay rise, union officials were urging an immediate return to work. Thus early in the strike two distinct positions – apart from that of the firm – emerged. It was in response to this situation that the RFSC was formed. The RFSC proceeded to express concern about democracy and the need for a substantial offer as a condition of a return to work. Once organised the RFSC saw itself not only as the prime protagonist of justice and as the genuine voice of the strikers, but also as a body possessed of some power and worthy of recognition: it became important to the RFSC that a settlement of the strike should involve concessions from the GMWU to this committee and its personnel. In addition to securing a substantial pay rise, a major objective of the RFSC was either to challenge the position of the union or its established officers: a third distinct position on the employees' side of the strike had developed.

From the point of view of Pilkingtons, the strike took on a course in which the firm was obliged to adopt the role of an almost passive but interested and injured by-stander. If it continued to be concerned with getting its factories back into production without conceding unnecessarily generous terms, there seemed to be little it could do to achieve that aim until after the 'parsons' poll'. If Pilkingtons gave some assistance to the union, it did not want to go too far in case events proved it necessary to talk with the RFSC.

In the closing stages of the strike a settlement involving all parties looked to be out of the question. The only solution seemed to lie in the eventual capitulation of one or other of the major parties for the point seemed to have passed where a compromise not involving a colossal loss of face could be agreed upon.

The strike was finally brought to an end after an am-

THE ESCALATION OF CONFLICT

biguous offer of mediation by an elder statesman of the trade union movement which quite by chance coincided with a weakening in RFSC morale. The terms of the TUC offer were such that they could be interpreted in different ways by the RFSC and the GMWU, although the offer in fact amounted to no more than the arrangement of a meeting.

This form of settlement naturally pleased Pilkingtons: it was keeping its hands clean and did not require the paying out of more money. No opposition was encountered from the rank and file either: since all the 'generals' were recommending a return to work it was obvious that the dispute was being called off on the best possible terms. The union was happy because it regarded the TUC intervention as a strategy for bringing the rebels to heel. (The loyal shop stewards went to the TUC believing that Feather was on their side.) Most of the RFSC members on the other hand saw the offer as a recognition that they represented a sizeable body of the Pilkington workers, and as a guarantee that the GMWU would make concessions. They too thought Feather was on their side. So the strike ended with everyone claiming victory. If no compromise was devised, a 'formula' was discovered – a formula that lent itself to different interpretations by different people with different interests.

Needless to say the formula solved nothing, indeed if anything it drove the RFSC and the GMWU further apart, but it did get an end to the strike and it did shift the struggle into another setting, on to other ground which, as it turned out, was more favourable to the GMWU.

One corrective seems to be necessary to what we have said in this section. If we seem to have given the impression that the strike was prolonged simply because of misunderstandings between the three main parties, and that a few well-timed manifestations of goodwill would have driven the parties into each others arms, then we have been misleading. If there had been a time when conciliatory gestures and more thoughtful tactics would have worked, it had passed by the middle of the third week. From then on the strike became a very straightforward power struggle. It was not so much that each party misunderstood the aims of the others, indeed the contrary, for the opposing generals understood each other very well. The point was that the interests of one cut right across the interests of the others. In such a situation of total war there were only two possible conclusions: capitulation, or an armistice arranged by the equivalent of the United Nations.

Strikes lasting for seven weeks are exceptional in Britain. It is possible however that the ever-changing processes of the Pilkington strike that transformed a spontaneous wildcat into a protracted struggle, may be present in other intractable disputes.

Some theorists of industrial disputes have suggested that the occurrence of strikes is likely to be least where a well-established trade union movement enjoys settled bargaining relations with employers. The trade unions have been regarded as institutions facilitating the settlement of grievances with the minimum of friction. No doubt there is much to be said for this theory, but our study also suggests that such arrangements can have the reverse effect of prolonging a strike once it is in motion.

If grievance settlement procedures create the means of resolving disputes peacefully, they simultaneously create, if crises continue to arise, opportunities for attempted coups d'etat or peasant rebellions simply because procedures may become ends in themselves rather than means to an end. There is then always a distinct possibility that an unofficial walk-out may escalate. This is a problem with which trade union leaders are constantly faced, and of which they will have become more aware following the trials of the GMWU in St Helens.

The prospects of peace

In recent years the public has been flooded with solutions to the 'strike problem'. The *Donovan Report*, the Labour Government's *In Place of Strife*, and the Conservative

THE PROSPECTS OF PEACE

Party's *Fair Deal at Work*, have all had a crack. But what does our study suggest? It suggests in the first place that none of the proposals in current circulation would have made any difference to the Pilkington strike.

To define strikes as a 'problem' is to imply they are abnormal. But the burden of our analysis of the Pilkington strike suggests that within the context of the contemporary industrial order strikes should be regarded as anything but extraordinary. A wildcat can break out in perfectly normal conditions, and the structure of the relations between employers, trade unions, governments and workers, guarantees that some strikes will grow from small beginnings to mighty struggles.

Many of the ideas of strikes, and the measures canvassed to prevent or curtail them, are based on the assumption that they are aberrations due to inept management or weak trade unions. Thus it is said on the one hand that if only management could be made more alert to the dangers of unrest, and if they could be prompted into overhauling consultative procedures all would flourish in a state of sweet harmony. It is said on the other hand, with regard to trade unions, that if they could be changed so as to exercise more control over shop stewards and to operate more efficient administrative machines, the possibility of conflict would be minimised.

During and after the Pilkington strike both management and union indulged in a great deal of introspection as to whether they could have done anything to prevent the strike. The GMWU wondered if a different branch structure could have made a difference, and management wondered if greater speed and vigour in sorting out the wages and negotiating systems might have helped. Well, would it?

We feel the answer is 'no'. Tinkering with wage levels and methods of payment, and lubricating negotiating procedures could not have *prevented* the strike – although it might have postponed it. The idea that the dispute could have been averted by managerial initiative is based upon the assumption that the strike was a response to definite and pressing problems: we have shown that it was not. The fact is that no

CONCLUSIONS

large firm with a necessarily complex organisation can completely banish friction from the shop-floor. Problems such as clerical errors cannot be organised out of existence. Working conditions, and the outlook of workers which is generated as a response in much of modern industry are such that 'incidents' are inevitable, and strikes in some cases inescapable. No amount of managerial ingenuity can remove what is built into the system.

Similar conclusions may be drawn with respect to the trade unions. By becoming stronger and more efficient organisations the unions may be able to take some of the steam out of some issues, but an equally possible development is that steam may be *injected*. The stronger and more professional a union's apparatus becomes the greater the likelihood of a disparity of outlook between officials and rank and file. The conditions that gave rise to a civil war in the GMWU during the Pilkington strike are likely to become more widespread as unions become more professional, more disciplined, and more a part of 'the establishment'.

If strikes cannot be avoided by the application of more organisational efficiency and goodwill either by management or trade unions, could they be legislated out of existence? Leading politicians of the Labour and Conservative Parties have looked to the law as a possible means of dealing with unofficial strikes. One legislative proposal has been that agreements should be made legally binding. The idea being that legally binding agreements would make unions careful to commit themselves only to terms with which members would be satisfied, and that it would impose on the unions an obligation to restrain their members from sundering agreements.

Such legislation would not have affected the Pilkington strike, indeed if anything it would probably have aggravated rather than prevented it. Since the strike was not a response to specific grievances it could not have been avoided by a carefully negotiated contract. Furthermore, if the union had been legally restrained from offering any succour to unofficial strikers, and if it had been under pressure actually to discipline its members, then a breach between rank and file and hierarchy would have been even more likely.

Other proposed reforms that would put obstacles in the way of union leaders displaying sympathy with unofficial strikers would probably have the same consequence. If trade unions were not legally permitted officially to call or recognise strikes without going through a process such as holding a secret ballot, it would be much more difficult to encapsulate localised disputes on the shop-floor. Placing legal restraints upon union officials offers little hope of reducing the incidence of strikes.

The most direct way of dealing with the unofficial striker would be to hang legal sanctions over his head. It has been suggested that a 'cooling-off' period or 'conciliation pause' should be imposed over an intended strike or on one that has actually started. Workers who failed to comply with such instructions to return to work and 'cool it' would expose themselves to penalties in the courts.

Against this it has been argued that it would be impracticable to start legal proceedings with large groups of strikers, and that the proceedings themselves would simply inflame strikers and delay a return to work. There is always the chance too that the law might be revealed as 'an ass'. These arguments are formidable.

Perhaps the knowledge that striking was in violation of the law would have made those Pilkington workers who were 'pressured' into striking more resistant. It is possible too that the threat of legal sanctions would have secured a return to work after the first weekend. On the other hand since the majority of the strikers were drawn into the dispute without carefully calculating the consequences, it is just as likely that the prospect of legal penalties would have exerted no deterrent effect whatsoever. As events turned out the strikers were effectively fined amounts up to £200 in lost earnings.

If it were possible to use the law to intimidate the potential striker then we could expect his attention to be drawn to other methods of protest, for the strike is only one weapon

CONCLUSIONS

in the arsenal: there remain the work-to-rule, the overtime ban, the go-slow, the sit-down strike, the sit-in or factory occupation, and sabotage. Would those who would like to legislate against the unofficial striker really like to encourage these alternatives?

Our enquiry suggests no simple remedy. Outbreaks of unrest, sometimes short-lived and sometimes of a longer duration, are elementary consequences of the contemporary economic structure. The nature of work, the terms of the employer–employee relationship, the integration of the trade unions into the power structure, all make strikes inevitable. Politicians, employers, and trade union officials may regard the strike, especially the wildcat, as a 'problem'. They may lecture strikers to the effect that bureaucratic management, bureaucratic trade unions, and desultory jobs are prices that have to paid for the benefits of industrialism. But there is another point of view, possibly only apparent to those who take part in strikes, that it is not they that are the problem; rather it is the economic structure, the trade unions, and the nature of work and authority.

A plausible lecture could be delivered to politicians, employers, and trade union leaders, containing the message that the 'nuisance' of the strike is part of the price *they* must pay if *they* want to reap the rewards of our economic system. Only when this point of view is appreciated will the reasons for striking be properly understood.

Appendix

SELECTED DOCUMENTS

BULLETIN FROM RANK AND FILE STRIKE COMMITTEE

13th May, 1970

The £12 Bribe

After six weeks on strike the NUGMW owes us £36 each. Last Sunday we decided overwhelmingly to get rid of the NUGMW. Within three hours they had announced that a 'hardship' allowance of £12 for men and £6 for women would be paid. *Not* weekly *strike pay* but a '*hardship*' *allowance* – payable, presumably, both to those who have been out for six weeks and the miserable handful who crept back a week ago – and then crept out again! SO WE SHOULD ASK THEM WHAT THEY HAVE DONE WITH THE OTHER £24!!! After all, it's *our* money – *we* paid it in.

Should we take the money?

We have already answered this question – it is *our* money – *we* paid it in over the years: so don't let the blacklegs paying *your* money out to *you* pretend they are being generous – and don't forget to ask them for the rest!!! And remember *they* cannot impose conditions when they pay *you your* money. But a word of warning – there may well be some of our brothers and sisters who may not be clear what they are signing – if so, be tolerant with them: they will have no intention of returning to work.

'Hardship Delegation' to London

Following the visit to the House of Commons by delegates from the Rank and File Committee in the company of our Member of Parliament, Leslie Spriggs, the following statement was agreed with Leslie Spriggs: *Quote*: The Deputation met the *Minister of State* at the Ministry of Health and Social Security, David Ennals, MP and Lord

APPENDIX

Collinson (Chairman, Supplementary Benefit Commission) to discuss the urgent need for financial help. The Minister of State heard our case and gave an undertaking to consider the need to set up a separate centre to deal with applicants individually. We are hopeful that each case of overriding urgency will receive more sympathetic consideration than they have in the past. *Unquote*.

Today's 'Pilkington Piffle'
Today's 'Pilkington Bulletin' again praises the NUGWU which must now top the poll as the bosses' favourite organisation. The laws on obscenity will not allow us to give the workers comments on this gallant band of men. The latest stunt is to threaten that if we don't all go back in the bosses' union we'll get the sack. They should try telling them that at Fords at Halewood – at the end of the day *it was the NUGMW that got the boot* – and not the men. And let's have less of this rubbish about lost orders etc: the only people who can put an end to the present situation are the DIRECTORS OF PILKINGTON BROTHERS – they will have to take *less profit* and pay more *wages*. The remedy is in their hands.

STOP PRESS Beware of Barbara Castle – she's got RED hair!

Quotes from the Press
'St Helens strikers are now calling it "the bosses' union" and are accusing its officials of having made "a mockery of trade unionism". The visiting union officials concurred with these views.' *Sunday Telegraph, 10th May*
'Thousands opted, at a mass meeting on Sunday, to leave their Union, the General and Municipal Workers, which will not declare the strike official.' *The Times, 12th May*
'... there is no disputing the feeling of genuine grievance on the lower deck which goes back some two years, and the widespread hostility to the union which refused to support the strike.' *The Times, 12th May*
'Mr Caughey, who has already refused police protection after a threat on his own life, said: "My wife is naturally

SELECTED DOCUMENTS

worried, but I feel we have the best protection in the world – the man-in-the-street is behind us."'

Lancs. Post & Chronicle, 12th May

. . .

The following lines are dedicated to Lord Cooper, General Secretary of the Nation Union of General and Municipal Workers.

(you will remember the NGMWU – it used to be the workers union at Pilkingtons. It finished on 10th May, 1970. RIP.)

Little Lord kneels at the foot of the bed
Looks underneath to see if there's a *Red*
Hush, hush, whisper who dares
Little Lord Cooper is saying his prayers.

God Bless Lord Harry – I know that's right
With a fellow Lord I never shall fight
For he's so right, and the men so wrong
Oh, God bless Bradburn and make him strong.

If I open my fingers a little bit more
I'll see Caughey's shadow behind the door
And if I lift my arms, and I'll throw more mud
Oh, God Bless Caughey and make him good.

I'll close down the office and stay in my bed
With fibreglass sheets right over my head
And keep wage agreements exceedingly small
And pretend to myself there's no strike at all.

Thank you God, for a lovely day
For long, long hours, and low, low pay
I've missed out something – now what can it be
Ah, now I remember it – GOD HELP ME!

14th May, 1970

GENERAL AND MUNICIPAL WORKERS' UNION

91 Branch Stewards

PILKINGTON DISPUTE
THE FACTS

What This Union Has Done

NEGOTIATED *two* agreements between October and April, raising lowest adult male rate from £12 12s 4d to £18 0s 0d.

NEGOTIATED the setting up of two working parties to deal with *your* grievances on (*a*) wages structure and (*b*) negotiating machinery.

NEGOTIATED the setting up of WORKS COMMITTEES to channel your views on pay structure and negotiating machinery to these working parties.

AGREED to an inquiry into 91 Branch.

NEGOTIATED hardship money of £12 for *all* full members and £6 for *all* half members. No conditions.

NEGOTIATED the setting up of a Court of Inquiry which will be in business next week. This will give the facts.

WHO ELSE HAS NEGOTIATED?
WHO ELSE CAN NEGOTIATE?
NO ONE ELSE
NO ONE ELSE CAN NEGOTIATE

NUGMW – SCAB UNION!

The National Union of General and Municipal Workers cannot be reformed – it can only be *replaced*.

* The Union is *undemocratic* – its officers are *appointed* – not *elected*.

* No attempt to change the union *from within* is tolerated. Even *criticism* can lead to expulsion. See 'Rules of the NUGWM' rule 43, paragraph 7.

SELECTED DOCUMENTS

* The stage has almost been reached of the leadership of the union being a 'family tradition' – thus Lord Cooper is related through his mother to Lord Dukes, the NUGMW General Secretary in the 30s.

* Lord Cooper also has many directorships: he is Governor of the London School of Business Studies; Director of Telefusion; Director of the National Ports Council; and also a Director of the *Atlas Foundation*, an organisation linked with the American Central Intelligence Agency (as was disclosed in the New York Herald Tribune on 15th August, 1967.)

* Nor is it surprising that Lord Cooper favours *Government fines* for unofficial strikers (see his evidence to the Royal Commission of Trade Unions on 5th July, 1966.)

* But Lord Cooper believes that the best way to keep workers in line is through 'representing' them in the NUGMW: he likes arrangements where the employer guarantees a 'closed shop' and in return the NUGMW promises to oppose unofficial action. *By 1966 the NUGMW had signed 315 such agreements.*

The Union's record at Fords is particularly revealing:

* in 1962 the NUGMW signed an agreement with the management to replace the power of the shop stewards with that of full time officials. Gradually, track speeds were increased by a third, and in the following year Ford produced 140,000 more vehicles with the same labour force.

* in the Ford strike of March 1969 the General and Municipal Workers Union, *alone out of ten unions*, refused to declare the strike official.

THE TRUTH IS OBVIOUS! The NUGMW is not a democratic union. It does not have elections. The rank and file have no control over its activities. It is AN AGENT OF THE EMPLOYERS both nationally and locally.

The question facing the men and women who have paid their money to this 'union' in the past is this: DOES THIS UNION DESERVE YOUR SUPPORT IN THE FUTURE?

The only long term solution is a set-to within the factories

APPENDIX

where the real power rests *with workers who work in the factories* and not with the over-paid, well fed gentlemen in the navy blue suits. The solution is in the hands of the workers *NOW!*

This is a Socialist Worker pamphlet. Meetings are held in the White Hart every Thursday.

PILKINGTON BROTHERS LIMITED   12th May, 1970
INDUSTRIAL DISPUTE
The Trading Position
The company must draw attention to the effect of the strike on the trading position.

In export markets many orders have already been lost and the efforts of the marketing sections have been squandered.

In the home market it is already clear that the biggest customer – the motor trade – will not again rely on Triplex alone for its supplies of safety glass. Many other home customers have placed large orders with our competitors – of whom there are plenty throughout the world – who are in a position to insist on long-term contracts.

There are several products which have been kept going at a loss because overall the Company has been profitable. One is sheet glass which provides many more jobs than any other form of glass making. Others are certain pressed, worked, and glass fibre products. Pilkington is not the only manufacturer of several of these products in the UK. The future of such products is increasingly at risk.

The damage already done to the Company's competitive position will take a long time to repair.

Loss of sales means inevitably that jobs are disappearing.

And it is not only the jobs of those on strike but jobs throughout the Company.

The questions every employee must ask himself or herself
1. Who is gaining from a continuation of the strike?
2. What possible advantages arise from continuing the

stoppage which cannot be gained after a return to work?

3. How seriously will prolonging the strike reduce the ability of the Company to ensure a high level of employment and real job security?

4. Where are jobs going to disappear because lost sales mean lost work? The unofficial strike committee has one thing to offer you – that many of you will have no jobs to come back to.

Eccleston Works will be the Works most immediately affected. Eccleston Works is an important customer of Cowley Hill and lower demand for Triplex products will have a serious effect on the float glass position.

Sheet Works will bear the brunt of lost exports – one million feet a week of sheet exports provides a very large number of jobs – in tanks, warehouses, timber yard, offices.

The surest way to ensure that jobs are lost is to damage your own best customers.

The Pilkington Joint Industrial Council

It is not the function of the Company to defend the position of any particular Union. No Union would want it to be so.

But for the record, The GMWU and the Company have an agreement under which an employee who ceases to be a member of the Union terminates his employment with the Company in any job which comes under the JIC.

Also for the record, last October's settlement between the Union and the Company at the JIC was for a 13½% increase in the wages bill.

In April, the same Union negotiated an increase of £3 for every JIC worker. The cost to the Company wage bill – a further 10%.

It has negotiated the Company's JIC wage bill up by nearly 25% since October last.

The Department of Employment & Productivity Gazette of February 1970 published Average Weekly Earnings for men 21 and over for a week in October 1969. Pilkington Glass JIC stood fourth in that table.

APPENDIX

That was BEFORE the application of the JIC agreement negotiated in September.

It was BEFORE the JIC negotiated £3 in April.

Together these make absolutely sure that Pilkington earnings stand comparison with industry throughout the country.

And they stand comparison on overtime worked. JIC average hours were 46·8 – the average for all industry was 46·5.

The Company is convinced that the £3 is a realistic settlement and it has no intention of improving on it before a return to work.

Another Union – a different JIC would not have achieved a significantly different settlement.

What Next?

The scenes outside Sheet Works last week will not be repeated. The Deputy Chief Constable of Lancashire has publicly stated that adequate resources are available to ensure safe conduct and law and order.

Where intimidation or threats are used you should report these immediately to the Town Police who will investigate every complaint brought to their notice.

The Company has welcomed the appointment of the Court of Inquiry and has undertaken to co-operate fully.

A continuation of this strike will make several things certain:

1. The development of an improved structure will be delayed.

2. It will not persuade the Company to amend its agreement either as to form or as to amount.

3. It will cause the Company to lay off a large number of employees as a decision whether to keep tanks on soak or put them out is forced on it.

4. The number of jobs which will disappear permanently will escalate rapidly from the number which have already been lost.

These effects cannot be in your real interests.

A recent bulletin of the Company advised employees wishing to return to work to await an announcement so that the arrangements to ensure unmolested access to and from the Works could be arranged.

The decision to return will be yours. But in order that there will be an orderly resumption of production you should get in touch with your elected Shop Steward or your Union official for guidance.

PILKINGTON BROTHERS LIMITED 6/GIR/M/T
26 April, 1960

TIME RECORDING REGULATIONS

The arrangements for time recording which are mentioned in Section 1(b) of the Works Rules and Regulations are as follows:

1. DAY WORKERS
(a) must clock in every morning and again after the dinner hour
(b) must not clock out when leaving their work at 12 noon Monday to Friday inclusive, unless finishing work for the day, but must clock out on all other occasions including those when they leave for dinner at any time other than 12 noon.
(c) must not leave their work without permission until the leaving off signal has been given.

2. SHIFT WORKERS
(a) must clock in when starting and clock out when finishing work.
(b) must not leave their work without permission unless relieved.

3. CLOCK CARDS
(a) each employee must clock only his or her own card
(b) cards must not be altered, folded, or defaced.
(c) cards must not be taken away from the appropriate racks at the clocking station concerned.

APPENDIX

4. PENALTIES

(a) *General*

Any employee who wilfully

(i) clocks another employee's card

(ii) arranges with another employee to clock his or her card

(iii) alters or defaces any clock card

will be liable to dismissal

(b) *Late Starting*

Employees will be considered late if they have not clocked in when the clock shows two minutes past the declared starting time. The penalty for being late will be a deduction of a quarter of an hour's time for each quarter of an hour or part of a quarter of an hour's lateness.

(c) *Early Finishing*

(i) DAY WORKERS who without permission leave their jobs whether working in or away from their Department before the three minutes hooter has sounded will be subject to the penalty of a deduction of a half hour's time. During the three minutes Employees may:

1. Return to their shops.
2. Put away their tools.
3. Wash their hands, get ready to go home, etc.

They must NOT:

1. Leave their department and proceed to the Lodges and Canteen.
2. Form queues at the Time Clocks.

(ii) SHIFT WORKERS who clock off without permission before the recognised finishing time will be subject to a deduction of a half hour's time for each half hour or part of half an hour lost.

(d) *Leaving The Works Without Permission*

Any employee who without permission leaves the Works before the recognised finishing time or in the case of a Shift Worker before he is relieved will be liable to dismissal.

PILKINGTON BROTHERS LIMITED

SPECIAL BULLETIN FROM RANK AND FILE COMMITTEE

Sunday 10/5/70

Brothers & Sisters,

We have now come to the parting of the ways. The NUGMW no longer serves any useful purpose for the workers of Pilkingtons. Therefore, arising from the general view expressed at last Sunday's meeting we are issuing the forms for contracting out of the NUGMW. By so doing we rob Pilkingtons of the last argument they have for not conceding our claim, that is, that the NUGMW think their offer 'fair and reasonable'.

. . . .

For a long time now the NUGMW has served as a vehicle for the promotion of ambitious individuals to positions of management. That is why it has been impossible to get certain shop stewards to reflect the views of the men they were supposed to represent. THEY are FINISHED now.

. . . .

(Quote from the *Manchester Guardian* on Monday, 4th May)

'The purpose of the union was clear from the start: to issue propaganda *against the strike leaders* and to erode the solidarity that had been a characteristic of the strike. *The union appeared to echo the employers' arguments throughout*, *refused* point blank to declare the strike *official*, and organised a ballot among Pilkington workers. The *result* of this ballot, however, vanished into thin air . . . although it is rumoured that *it showed a big majority in favour of the strike*.'

. . . .

The latest gimmick is the appeal by the firm to return to work while the Court of Inquiry set up by the DEP looks into the affair. What is Pilkington saying now? That the Court cannot look into this whole shameful affair unless we

APPENDIX

are actually back in the factories making glass? Remember, a Court of Inquiry is *not* an Arbitration Tribunal – it is simply a body of men who look into the *cause* of the strike. So let us tell them now – as if they don't know: the cause of this strike was thirty years of too much work for too little pay. Let *them* get on with their *Inquiry* while we get on with our *strike*.

BULLETIN No 1

ISSUED BY THE RANK AND FILE STRIKE COMMITTEE

A changed situation

The situation in St Helens has changed dramatically in the past few days. *Pilkington Brothers have now come out in their true colours*. The bribes have failed to secure a return to work. The lies told to the Press and television have not worked. The people of St Helens have not been deceived. The pressure put on the widows and pensioners has been of no avail. *PILKINGTONS ARE FRIGHTENED*.

Why are they scared?

They are scared because the glass workers of St Helens have woken at last from half a century of drugged sleep. DRUGGED BY LONG HOURS OF WORK FOR MISERABLE WAGES. Drugged by the argument that the firm was a great benefactor doing the workers a favour by siting itself in St Helens. DRUGGED BY THE NICE HAPPY PICTURE OF A KIND HEARTED 'FAMILY CONCERN' WITH AN ECCENTRIC MILLIONAIRE RIDING TO WORK ON A BIKE!

Now the mask is off

From: *The Pilkington Affair 'The Guardian' Monday, 4th May*. Quote: '... the company decided to pay to all its other workers the £3 a week rejected by the strikers. Pilkington would certainly refute any suggestion of *strike-breaking* but this is what the company *did* as effectively as if it had brought in substitute workers.'

SELECTED DOCUMENTS

What of the 'union'?
Another quote from 'The Guardian'. 'The purpose ... of the union ... was clear from the start: to issue propaganda *against the strike leaders* and to erode the solidarity that had been a characteristic of the strike.' *The Guardian* 4/5/70. Of the NUGMW Geoffrey Whiteley wrote on the business page of *The Guardian*: 'The union ... appeared to echo the employers' arguments throughout, refused point-blank to declare the strike official, and organised a ballot among Pilkington workers. *The result of this* ballot, however, appears to have vanished into thin air – although it is ... rumoured that it showed a big majority in favour of the strike.'

Back to the Thirties
Now the tactics employed by the meanest firm in Britain against our fathers, our grandfathers, and our great-grandfathers are being reintroduced. *Policemen on horseback, policemen from Preston, Bolton, and other parts of Lancashire* ... brought into St Helens so that Lord Harry and Lord Cooper can have their way. And to hell with democracy as far as they're concerned!!!

We are not alone
A huge campaign of support is now under way. This support comes first of all from the ordinary people of St Helens. It comes from the building workers, the TGWU, the AEF, and is now growing in volume from the Merseyside and beyond. Pilkingtons are now planning a semi-military operation to get a miserable handful of miserable men back into the factories. The fact that such men will be sent to Coventry, will be boycotted and shunned by all decent people *for the rest of their natural lives* means nothing to Lord Pilkington.

Discipline and determination are needed
In the next few days we shall be subjected to tremendous pressures. We still hope to avert violence *because that is not part of our policy*. But the threats being uttered by the firm

APPENDIX

still produce one answer from us – £5 NOW and *the rest* later!

Dear Sir and Brother,

Enclosed find £8/10/- a donation to your strike fund. We have had our own experience of being deserted by our Union and being pilloried by the press and know how you feel. But we are sure you will win in the end, and give strength to others in actions against their employers and reactionary Union leaders.

Hoping for an early and successful end to your fight.

Yours fraternally,

Dear Brother,

I have pleasure in sending you £10 – the result of a collection in the ... branch of IS in aid of your strike fund.

Your courage in staying out for seven weeks in the face of formidable opposition from the company, the GMWU, the press and the government is much admired down here in London.

We all hope that you win your just fight.

Yours fraternally,

WISHING YOU SUCCESS IN YOUR PRESENT DISPUTE STOP LETTER AND CONTRIBUTION FOLLOWING = ...JOINT SHOP STEWARD COMMITTEE SHEFFIELD

From a branch of the National Union of Vehicle Builders:

Dear Brother,

Many thanks for sending me a collection sheet. I am sure many of my friends will desire to contribute to your fund and wish you and your brothers success in your struggle against Pilkington and the GMWU.

I have passed your address on to several Branch Secretaries and Chairmen who will, I am sure, be forwarding support also. As soon as I have contacted all my colleagues I will forward contributions.

My personal wishes for success in the fight and fraternal greetings –

From a London Tenants Association:

Dear Brother,

I enclose a cheque for £15 for your strike. You would have had this earlier but we could not find out where to send it.

I also enclose a pamphlet which describes with which this Association and others in London have been connected with – the GLC rent battle. We too found the need to have a completely new organisation to fight our rent battles, the old organisations of the Labour movement having joined the establishment.

Best of luck, fraternally,

Dear Brother,

I have pleasure in sending you a cheque for £150 on behalf of the ... Plant Stewards Committee (Ford Halewood), as a donation towards your fighting fund. May I say on behalf of all of us at Fords that our hearts are with you in your struggle and may a resounding and conclusive victory be yours. Again, best wishes from the ... and may this new association between our two organisations go from strength to strength.

Yours fraternally,

Dear Mr ...

Thanks for the receipt and the letter that came with it. Particularly the letter, and I send you another small donation before I send it to some other cause!

My striking days are over now but I recall my first strike many years ago was a union recognition strike and the union that sent its members into work during that strike was the GMWU.

APPENDIX

Later I left the workshop floor and escaped into higher levels of management and there, of course, came across, among other unions, the GMWU. I came to despise that union because of its pro-management and anti-union attitudes. They were highly regarded by the managements because they always said the sort of things the management wanted to hear. In particular I came across the late Jim Matthews. There was not a dirty piece of intrigue too dirty for him to be involved in. He was then associated with some enquiry organisation of some sort and would seek out past histories of workers and hand on the information to the employers.

On one particular occasion when negotiations about a money increase were under way at national level he actually hung back as an adjournment took place and whispered advice to the employers representatives, of whom I was one and present. 'Don't give it to them,' were his actual words. To me Jim Matthews was, and although now dead, still is the true image of the GMWU.

I wish you well in the work you have undertaken. I have nothing but admiration for the splendid struggle you and your colleagues have put up.

Yours sincerely

From a branch of the National Union of Mineworkers:

Dear Brother,

Enclosed is a cheque for £12. 7. 0 being a donation to your Strike Fund. We are only a small branch so I think you will appreciate the problems, we do however wish you all the best in your struggle.

Yours faithfully,

From AEF Shop Stewards in Wolverhampton:

Dear Sir,

We have not received an appeal from you in reference to your strike, but due to an article in *The Morning Star* we obtained the address.

SELECTED DOCUMENTS

A shop floor meeting was held and it was proposed that we donate a sum of £25 to your fund in an endeavour to see justice done and to help in this fight in times of dire need. Our shop floor is composed of roughly 100 men who heartily support you in your battle ... The whole of our department is made up of AEF members but we too would like to see justice carried out in the name of SOCIALISM no matter what Union, Class or Creed.

Yours fraternally,

From a Cambridge College:

Dear Sir,

Please accept the enclosed £8 cheque, it being the result of a collection taken here. We support you in your struggles against all opposition; not even Lord Cooper can make wildcats into mice.

Yours fraternally,

From a London branch of the Association of Scientific, Technical and Managerial Staffs:

Dear Sir and Brother,

My branch has asked me to send you its support in your struggle to have your demands met, and to obtain the sort of organisation which recognises the requirements and interests of the workers at Pilkingtons. It sends a small donation ...

Yours fraternally,

Dear Sir,

Please accept the enclosed for your strikers. Pilkingtons threatening to sack all those who changed their Union, was for me the last straw. What a bloody nerve!

Well Wisher.

From a Pilkington pensioner:

Dear Sir,

Please accept £1 enclosed. It is not much I know, but it's effort that wins through. I am a pensioner of 84 years and

APPENDIX

spent 52 years of it with Pilkingtons, and may I say that this strike has come 40 years too late. In my working days it was on the cards several times, but we did not seem to get united as at present. Keep united on this issue, this will win the day. This as I see it goes deeper than £3 on or off the basic wage. Pilkington has held the whip hand for too long.

Good luck to all the committee and workers.

From an Amalgamated Society of Woodworkers' shop steward:

Dear Brother,

Thank you for your appeal form for finance in your very just fight for a reasonable wage ... I am sure your great fight against injustice will be an inspiration to many more underpaid over-exploited industrial workers.

Despite the lies, half truths from the establishment media of press and television, and of course corrupt trade union organisations.

I enclose a few of Jack London's *Description of a Scab*. I have found in the past they have helped to shame the weak element from passing over the picket line.

Best wishes,
Yours in unity,

From a docker's wife:

Dear Brother,

Enclosing 5/- donation and a few stamps towards your strike fund – with very best wishes for your victory.

Take no notice of the mudslingers. The working man has to make a stand in the best way he knows how – for justice and a living wage. Nobody has had more mud slung at them than the Liverpool dockers. I am married to one and I know. He can no longer work because his health broke down through working overtime year in and year out.

Good luck to all your lads and lasses. Don't waste a stamp sending a receipt.

Yours sincerely,

SELECTED DOCUMENTS

Dear Brother,

Enclosed a cheque for £10-2-6 for the strike fund. This has been collected by Bristol University student socialists since we saw your address in *Socialist Worker* ...

Hope it will be of some use,

Yours fraternally,

From Cardiff University students:

Dear Comrade,

We enclose £2 towards the strike fund – best of luck with your strike demands. Don't consider going back as long as the workers in Pontypool are kept from their work.

From a Trades Council in the NW:

Dear Brother,

The above council is enclosing a donation of £10 to your strike fund, and in doing so wish you success in your struggle for better wages and conditions.

The trades council is very much concerned at the actions of the mass media in their biased coverage of men and women who are taking action in defence of their living standards. Your struggle and the justice of your cause deserves the support of the whole Labour Movement, and the trades council has already circulated all its affiliated bodies to this effect on your behalf.

Fraternal greetings to your committee on behalf of our trades council, and best wishes for a successful end and victory in your dispute.

Yours fraternally,

From Leeds Clothing Workers:

Dear Sir,

Please find enclosed £10 hoping it will be of some assistance to your Fellow Workers who are suffering hardship.

We the ... are organising collecting sheets to be passed

APPENDIX

round all clothing factories in Leeds and district. As this will take time, it may be 3 to 4 weeks before you will hear any word from us ...

We all here in the clothing industry wish you the best in your struggle for better wages.

Yours sincerely,

Dear Fellow workers,

Please accept this small donation (10/-) towards your strike fund. We are a working class family – Father AEF shop steward, son young socialist and worker, daughter telephonist, mother launderette worker and nine year old son who is learning about the establishment fast. We all wish you the very best of luck and do not believe the smears that have been levelled at you all.

Keep fighting and may *you win*.

The ... family.

Fontana Books

Fontana is at present best known (outside the field of popular fiction) for its extensive lists of books on history, philosophy, and theology.

Now, however, the list is expanding rapidly to include most main subjects. Now series, sometimes extensive series, of books are being specially commissioned in most main subjects – on literature, politics, economics, education, geography, sociology, psychology, and others. At the same time, the number of paperback reprints of books published in hardcover editions is being increased.

Further information on Fontana's present list and future plans can be obtained from:

The Non-Fiction Editor,
Fontana Books,
14 St. James's Place,
London, S.W.1.

Fontana History

Fontana History includes the well-known History of Europe, edited by J. H. Plumb, and the Fontana Economic History of Europe, edited by Carlo Cipolla. Four new series are in preparation. Books now available include:

The Practice of History G. R. Elton

Debates with Historians Peter Geyl

Domesday Book and Beyond F. W. Maitland

The English Reformation A. G. Dickens

The Nation State and Self-Determination Alfred Cobban

Europe and The French Revolution Albert Sorel

Russia 1917: The February Revolution George Katkov

The Downfall of the Liberal Party Trevor Wilson

The Trial of Charles I C. V. Wedgwood

The King's Peace 1637–1641 C. V. Wedgwood

The King's War 1641–1647 C. V. Wedgwood

Fontana Literature

Literature may well provide the largest single section of the expanding Fontana list. In preparation is an extensive critical history of English, American, and Commonwealth literature, and a series on literature and sociology. Books now available include:

Axel's Castle Edmund Wilson

Sartre Iris Murdoch

The Brontë Story Margaret Lane

Early Victorian Novelists David Cecil

The Stricken Deer David Cecil

Modern Australian Writing Edited by Geoffrey Dutton

Modern Poets on Modern Poetry Edited by James Scully

Fontana Politics

The first two of what will be an extensive series of original books appeared in 1970. They are:

Governing Britain A. H. Hanson and Malcolm Walles

The Commons in Transition Edited by A. H. Hanson and Bernard Crick

Other books now available include:

The English Constitution Walter Bagehot
Edited by R. H. S. Crossman

Asquith Roy Jenkins

Sir Charles Dilke Roy Jenkins

Marx and Engels: Basic Writings
Edited by Lewis S. Feuer

Democracy in America de Tocqueville
Edited by J. P. Mayer and Max Lerner

The Downfall of the Liberal Party 1914–1935
Trevor Wilson

War and Modern Society Alastair Buchan

Fontana Social Sciences

An extensive economics series begins publication in Spring 1971, with up to seven volumes. A sociology series is in preparation, together with a series combining sociology and literature, and sociology and history. Other books available include:

The Sociology of Modern Britain
Edited by Eric Butterworth and David Weir

People and Cities Stephen Verney

The Acquisitive Society R. H. Tawney

Memories, Dreams, Reflections C. J. Jung

African Genesis Robert Ardrey

The Territorial Imperative Robert Ardrey

The Varieties of Religious Experience William James

Lectures on Economic Principles Sir Dennis Robertson

Essays in Money and Interest Sir Dennis Robertson

Fontana Modern Masters

This series provides authoritative and critical introductions to the most influential and seminal minds of our time. Books already published include:

Camus Conor Cruise O'Brien
Chomsky John Lyons
Fanon David Caute
Guevara Andrew Sinclair
Joyce John Gross
Lévi-Strauss Edmund Leach
Lukács George Lichtheim
Marcuse Alasdair MacIntyre
McLuhan Jonathan Miller
Orwell Raymond Williams
Wittgenstein David Pears

'We have here, in fact, the beginnings of what promises to be an important publishing enterprise. This series is just what is needed by the so-called "general reader" in search of a guide to intellectual currents that clash so confusingly in a confused world.'

Times Literary Supplement

Many more are in preparation including:

Buckminster Fuller Allan Temko
Eliot Stephen Spender
Freud Richard Wollheim
Gandhi George Woodcock
D. H. Lawrence Frank Kermode
Lenin Robert Conquest
Mailer Richard Poirier
Reich Charles Rycroft
Russell A. J. Ayer
Trotsky Philip Rahv
Yeats Denis Donoghue